Springer Theses

Recognizing Outstanding Ph.D. Research

Aims and Scope

The series "Springer Theses" brings together a selection of the very best Ph.D. theses from around the world and across the physical sciences. Nominated and endorsed by two recognized specialists, each published volume has been selected for its scientific excellence and the high impact of its contents for the pertinent field of research. For greater accessibility to non-specialists, the published versions include an extended introduction, as well as a foreword by the student's supervisor explaining the special relevance of the work for the field. As a whole, the series will provide a valuable resource both for newcomers to the research fields described, and for other scientists seeking detailed background information on special questions. Finally, it provides an accredited documentation of the valuable contributions made by today's younger generation of scientists.

Theses are accepted into the series by invited nomination only and must fulfill all of the following criteria

- They must be written in good English.
- The topic should fall within the confines of Chemistry, Physics, Earth Sciences, Engineering and related interdisciplinary fields such as Materials, Nanoscience, Chemical Engineering, Complex Systems and Biophysics.
- The work reported in the thesis must represent a significant scientific advance.
- If the thesis includes previously published material, permission to reproduce this must be gained from the respective copyright holder.
- They must have been examined and passed during the 12 months prior to nomination.
- Each thesis should include a foreword by the supervisor outlining the significance of its content.
- The theses should have a clearly defined structure including an introduction accessible to scientists not expert in that particular field.

More information about this series at http://www.springer.com/series/8790

Alba Fernández Barral

Extreme Particle Acceleration in Microquasar Jets and Pulsar Wind Nebulae with the MAGIC Telescopes

Doctoral Thesis accepted by
the Universitat Autònoma de Barcelona, Bellaterra, Spain

 Springer

Author
Dr. Alba Fernández Barral
CTAO gGmbH (Headquarters)
Heidelberg, Germany

and

CTAO gGmbH (Local Office)
Bologna, Italy

and

IFAE–BIST
Edifici Cn, Universitat Autònoma de
  Barcelona
Bellaterra (Barcelona), Spain

Supervisor
Dr. Oscar Blanch Bigas
IFAE–BIST
Edifici Cn, Universitat Autònoma
  de Barcelona
Bellaterra (Barcelona), Spain

ISSN 2190-5053   ISSN 2190-5061   (electronic)
Springer Theses
ISBN 978-3-030-07371-8   ISBN 978-3-319-97538-2   (eBook)
https://doi.org/10.1007/978-3-319-97538-2

This Springer imprint is published by the registered company Springer Nature Switzerland AG
The registered company address is: Gewerbestrasse 11, 6330 Cham, Switzerland

Allà ets l'essència de tot. Pels crítics, els paradigmes i els cànons pels que es regien estan obsolets de cop. Que amb l'aura que tu desprens s'imposa, de forma urgent, fer una revisió dels clàssics. Que ets jove eternament.

Els amics de les arts
Museu d'Història Natural

A mis padres, a BRAIS.
Os pertenece cada palabra aquí escrita.

Supervisor's Foreword

The birth of gamma-ray astronomy, or astroparticle physics, can be dated back to 1989, when the Crab Nebula was first observed at TeV energies with a ground-based imaging atmospheric Cherenkov telescope. In less than 30 years, we have witnessed many great discoveries and technological advances in this new field. Dr. Alba Fernández Barral's thesis adds to this step-by-step evolution, providing not only outstanding physical results but also aiding the development of the next generation of Cherenkov telescopes, validating and characterizing several subsystems of the telescope's camera.

In order to contribute her two pennyworth to the understanding of different astrophysical phenomena, the author deepened in the performance, hardware, and software of the MAGIC telescopes, located in the Canary Island of La Palma. She also carried out several scientific sojourns to expand her knowledge of different gamma-ray detection techniques, such as those used by *Fermi*-LAT and HAWC, which observe the sky at complementary energy bands alongside MAGIC.

Mainly focused on the physics behind the relativistic jets displayed by microquasars, her work provides a complete view of Cygnus X-1 in the GeV–TeV band. This source, one of the most famous and well-studied binary systems, lacked of conclusive information regarding gamma-ray emission processes and relied on theoretical hypotheses. With her work using both *Fermi*-LAT and MAGIC data, the author was able to not only detect the source for the first time in the gamma-ray band, but also provide constraints on the production site and mechanism that gives rise to gamma rays within the jets, pointing to inverse Compton on stellar photons as the responsible process. This allows to limit and even refute different theories, such as the so-called *advection-dominated accretion flow* models. Although not detected at very high energies, a deep study on Cygnus X-3 and V404 Cygni is also given, providing a state-of-the-art overview of the best microquasar candidates to emit gamma rays during exceptional transient jet events.

A population study of pulsar wind nebulae (PWNe) was recently released after the sky scan performed by the H.E.S.S. Collaboration at the inner parts of the Milky Way. Under the premise that this type of source is the most numerous in our galaxy, the author coordinated a project analyzing and discussing the data for five optimal candidates formerly observed by MAGIC, in order to shed light on the particle acceleration of PWNe at the outer parts of our galaxy. Moreover, while working with HAWC groups, she meticulously investigated the best candidates for MAGIC among the new TeV sources detected and published by HAWC. She led and carried out a joint MAGIC, HAWC, and *Fermi*-LAT project in which the alleged PWNe nature of three new sources is deeply discussed and the size of each of them constrained.

Besides the quantity and quality of her work, it is worth stressing Alba's involvement on getting those results: She was always leading the analysis, interpretation, and edition as well as coordinating the projects she worked on.

Barcelona, Spain   Dr. Oscar Blanch Bigas
July 2018

Preface

Throughout our entire history, we humans have strived to unravel the mysteries with which the deep universe challenges us. In our humble beginnings, this task was performed with our naked eyes, by gazing at the stars and planets and wondering how far away they were and how they moved in the night sky. For many centuries, only the visible universe was reachable for us, but extraordinary achievements were accomplished despite the limited tools: We discovered, for example, that our planet was not the center of the universe, owing to Nicolaus Copernicus' observations and his heliocentric model. From Copernicus' epoch up to now, the development of new technologies and the advancement of our own understanding of the Cosmos allowed us to disentangle many riddles. Fortunately, this natural curiosity that leads us to improve never ends, and we face new questions that challenge our capacity as scientists. In the present thesis, I focus on a small fraction of this science: the gamma-ray astronomy. The bulk of results presented here was obtained by means of the imaging atmospheric Cherenkov technique, with the so-called MAGIC telescopes. Once arrived at the earth, a gamma ray interacts with atmospheric nuclei producing charged particles. The latter can move faster than the speed of light in air, giving rise to a very fast bluish light, known as Cherenkov light. The imaging atmospheric Cherenkov technique is based on the detection of that light, from which we can obtain information about the primary gamma ray. Other gamma-ray detection techniques included in the thesis, such as satellites or water Cherenkov tanks, have also been proved to play an important role to disentangle the gamma-ray sky. All these techniques are energetically complementary to each other, helping us to understand deeply the processes that take place on different astrophysical sources. This thesis comprises my efforts on the study of particle acceleration and gamma-ray production mechanisms inside galactic relativistic jets displayed by microquasars and the shocks produced in pulsar wind nebulae (PWNe), using in several cases all the aforementioned detection techniques.

Part I of the thesis encompasses an introduction to non-thermal universe as well as a detailed description of different gamma-ray detectors. First, I delve into the production and absorption mechanisms that govern the gamma-ray emission, along with the sources that may give rise to these gamma rays. I also provide a general overview of the imaging atmospheric Cherenkov technique and a very extended description of the hardware and data analysis procedure of the MAGIC system, for all the different performances that the telescopes underwent along the years. A description of Cherenkov Telescope Array (CTA), the future, and currently under construction, generation of Cherenkov telescopes, is also included. Moreover, I give an introduction to the HAWC Observatory and the *Fermi*-LAT satellite, including a brief description of the performance and analysis of the latter, as results from both instruments are used in the discussion of galactic sources shown in this thesis.

The scientific part of the thesis is covered in Parts II and III. In the former, I focus on the so-called microquasars, binary systems that show relativistic outflows at certain accretion states. These objects are optimal laboratories for the study of particle acceleration inside relativistic jets and the resulting transient or steady gamma-ray emission on accessible timescales. After giving an extended overview of these systems, I discuss results from three of the best microquasar candidates to emit very high energy (VHE) gamma rays: Cygnus X-1, Cygnus X-3, and V404 Cygni. I investigate Cygnus X-1 making use of publicly available *Fermi*-LAT data, which leads to the first detection of the system in the high energy (HE) regime. This also constitutes the first firmly gamma-ray detection of a black hole (BH) binary system. This result allows me to constrain the gamma-ray emission mechanism and the production site within the relativistic jets ejected by the source. This study is complete at higher energies with MAGIC data obtained during a long-term campaign between 2007 and 2014. The thesis also contains the VHE results of Cygnus X-3 under a major flaring period observed by MAGIC in 2016. MAGIC performed a follow-up of the activity of the source from the very beginning of the flux increase (in radio and HE) until the cease of the outburst 1 month later, owing to a dedicated trigger strategy started in 2013. I report the most constraining gamma-ray flux limits obtained in the VHE band and discuss the corresponding implications. I also include the results of the analysis of the low-mass microquasar V404 Cygni during its flaring activity in June 2015, after 25 years in quiescent state. I provide an estimation of the gamma-ray opacity, and pointing to low particle acceleration rate inside the jets of the system or not enough energetics of the VHE emitter. On the other side, Part III is focused on the study of PWNe. I analyze five PWNe candidates and set the results in the context of the TeV PWN population study performed by the High Energy Stereoscopic System (H.E.S.S.) Collaboration. Along with these results, I discuss the importance of the target photon field together with characteristic features of the pulsars hosted by these PWNe to emit gamma rays. In this thesis, I also present the first joint work between the MAGIC telescopes, the HAWC Observatory, and the *Fermi*-LAT satellite, which opens the door to future synergy projects. Here, I show the analysis of MAGIC data on two new TeV sources detected by HAWC, giving constraints on their size and discussing in detail their hypothetical PWN nature. Owing to the contextualization of these two candidates among the known TeV PWNe population and a study of the parent population that produces the TeV gamma-ray emission, I discard these new sources to be PWNe.

Scientific results outside these two topics are also included in Appendix A. With the aim of providing information about the density profile of the ejected material, I show the first and only results at VHE on a Type Ia Supernova, the so-called SN 2014J, the closest supernova in decades.

In Part IV, I present the technical work performed during my thesis for the future CTA instrument, focused on the camera hardware for the large-size telescope (LST). I first present a full overview of its different parts, along with the quality control (QC) tests for several subsystems, among which the power supplies and trigger mezzanines stand out.

All these scientific and technical results are summarized at the end in a conclusion chapter.

Bologna, Italy   Dr. Alba Fernández Barral

Declaration

I herewith declare that I have produced this thesis without the prohibited assistance of third parties and without making use of aids other than those specified; notions taken over directly or indirectly from other sources have been identified as such. This paper has not previously been presented in identical or similar form to any other Spanish or foreign examination board.

The thesis work was conducted from September 2013 to October 2017 under the supervision of Oscar Blanch Bigas at Barcelona.

Barcelona, Spain
October 2017

Acknowledgements

Xa quedou máis do que queda. Con estas palabras me animaba siempre mi padre cuando la tensión, el cansancio o el agobio se apoderaban de mí y creía que no iba a ser capaz de dar un poco más. Pero lo fui, papá. Y ahora, ya no queda nada. Esta etapa ha llegado a su fin.

Es difícil agradecer en pocas líneas a toda la gente que ha estado a mi lado durante este proceso. Quizás lo más justo es mirar hacia atrás y agradecer a todos los que estuvisteis apoyándome a lo largo de todo el camino.

Todo comenzó, sin lugar a dudas, en una academia de Betanzos, donde mi tío Sito hacía fácil lo que en aquel momento parecía tan complicado. Tío Sito, aunque nunca te lo dije, este amor por la física comenzó contigo.

En mi viaje hacia la vida universitaria, y durante los primeros años de ésta, tuve que vivir una etapa negra. Pero de todo se aprende en esta vida, y ahora sé que esos años me hicieron más fuerte y me ayudaron a valorar la felicidad y a las personas. Personas como Antonio y Miguel, que estuvieron siempre a mi lado. Gracias de todo corazón por haber estado ahí, por haberme acompañado esos años, por ser mis amigos.

En la universidad me encontré con el mejor grupo de actores que la física puede dar. Juan, Mónica, Tamara, María, Aida, Díaz, y tú también Miguel, tranquilos, si no triunfamos en la física, podemos vivir de la interpretación! Entre las aulas de Santiago, tuve también el gran placer de hacerme amiga del dúo inseparable, Víctor e Izan. Espero siempre con impaciencia nuestras reuniones anuales frente a un café, una copa o un cacaolat para ponernos al día sobre nuestras vidas. Y en estas líneas tienen que estar aquellos junto a quienes compartí tantas horas en la facultad o en "la Conchi", bailes en Retablo, cenas en casa y muchos muchos cafés (o mentapoleas con hielo y sacarina) en la cafetería de la facultad. Gara, Pau, Diego, Sergio, siempre seréis *mis reyes* de Santiago. En este recuerdo a Santiago, no puede faltar mis más sincero agradecimiento a José Edelstein y Juan Antonio Garzón Heydt por la confianza que depositaron en mí. José, Hans, sin vuestro apoyo no habría podido escribir esta tesis.

Y desde Santiago llegé a Barcelona, donde empezaría el máster en la UB. Ahí tuve la suerte de empezar a adentrarme en el mundo de la física de muy altas energías de la mano de Josep María Paredes, Marc Ribó y Valentí Bosch-Ramón, de los cuales seguiría aprendiendo durante el doctorado. También tengo que agradecerle a Andreu Sanuy su paciencia en el laboratorio. Gracias a toda la gente del máster que hicisteis de ése, un año maravilloso. Pero por supuesto, tengo que destacar a los primeros *master and commanders*: Andrea, Víctor, Alfonso (El Guapo), Ari y Adri.

En febrero del 2013 se me abrirían las puertas de IFAE, donde comenzaría el gran viaje del doctorado un 1 de septiembre. Sonrío al pensar en toda la gente que apareció en esta etapa de mi vida, a la que tengo tanto que agradecer.

Pero, como no puede ser de otra manera, empezaré con mi supervisor, con quien he tenido el enorme placer de haber trabajado estos cuatro años. Oscar, muchísimas gracias por todo lo que me has enseñado, por tu paciencia y dedicación, por tu empatía, por no poner nunca mala cara ante una duda y hacer siempre un hueco para discutir nuestros mil proyectos. No podría haber tenido un mejor director, así que espero de verdad haber estado a la altura como estudiante.

A Javi quiero agradecerle toda la ayuda que me brindó cuando tenía un problema, aunque me delatase como *almost senior student*. A Abelardo quiero darle las gracias por ser tan intenso y realista en los group meetings, una forma maravillosa de aprender lo bueno y lo malo de este campo. A Juan-SeniorPrize 2015 le agradezco que nos dijese sin tapujos cuando "it is all wrong", incitándonos así a mejorar. Al Manel li vull agrair les seves lliçons de hardware, durant les que sempre vaig aprendre alguna cosa nova. A Matteo y Ramón les quiero agradecer su ayuda como directores de IFAE y a Enrique como tutor.

En IFAE he compartido despacho, ideas y dudas, cafés y chupitos, risas y lágrimas con gente maravillosa que ha hecho de estos años una etapa absolutamente inolvidable. Comienzo con los postdocs que van y vienen de IFAE, pero a los que yo siempre recuerdo con mucho cariño. Julian, I do not have enough lines here to thank you for all your help, always with a smile in your face. JohnE, *qué pasa contigo, tío?*, I want to thank you for all the talks we had, with the future (or should I say reality?) as the main topic. Paolo, el domingo vermouth con Dani y Quique a las 12. A parte de eso, gracias por tu entusiasmo, por no fallar nunca a un plan y por haber recorrido Barcelona a pie y metro con una silla al lado. Tarek, a ti y a tus políticamente incorrectas frases ya os conocía, pero voy a echar de menos que alguien toque a la puerta de mi despacho tan característicamente para tomar el café de la mañana. Scott, thank you very much for all your help in the lab and everything I learned during our talks. Koji, trips did not allow me to work with you too much, but I hope your stay at IFAE is as good as it was mine.

Y en dos despachos unidos por una puerta que raramente se cierra, he conocido gente a la que llevaré siempre en el corazón. Adiv, mi compañero de enfrente durante dos años, gracias por todos los paseos domingueros por Barcelona y por las sesiones de fotos. Leyre, la meva companya de català, gracias por tu dulzura y ser la mejor compañera de viajes que podía tener, *tronca*!. Daniele, Merve and Elia, it was a pleasure to spend time with you at IFAE, I hope you follow our old traditions in the office. And the *special ones*: Dani, Quim, con vosotros comencé el doctorado y con vosotros lo voy a acabar. Ha sido un enorme placer teneros de compañeros y, por supuesto, de amigos. Dani, gracias por brindarnos tranquilidad y parsimonia, excepto cuando hay un "ashuwea" de por medio. Quim, el mejor compañero de despacho que podría haber tenido. Echaré mucho de menos nuestras discusiones de pizarra y mandarnos vídeos que nos alegran la hora del café. Gracias también al MAGIC-like group, más conocido por DES: Carlos, Pauline, Judit, Alex, Marco, y también Bernat, un placer haber compartido estos años y cafés con vosotros. Y a Jelena, por aguantar esos cafés al otro lado de la pared.

A los ingenieros de IFAE, Oscar Martínez, Juan Boix, Pepe Illa, Oscar Abril, gracias por vuestra ayuda y dedicación todos estos años. Y cómo no, gracias a Cristina por mantenernos siempre al tanto de todo y sobre todo por preocuparse de nosotros. Els teus nens es fan grans...

Y así pasó mi tiempo en IFAE, entre trabajo y amigos. Entre un teléfono falso en Madrid, la azotea de un piso en Roma desde donde se veía toda la ciudad, el inolvidable viaje a Florencia, el vino encima de un muro imposible de escalar, el pueblo perdido de Bulgaria donde el dueño de la casa hablaba "nuestro" idioma (*Ciao, come stai?*), Sofia y el fatídico destino de la furgoneta (a pesar de la alarma que indicaba que todo iba bien), PLOVEdiv y su flan, y su increíble bar en el segundo piso donde regalan sombreros de pirata, el karaoke de Berlín donde Twist and Shout se convirtió en nuestra canción, los shifts que coinciden con Los Indianos. Y cómo no, los vermouths, los mojitos, las mejores bravas de Barcelona aunque "ya no las hacen como antes", el bus 11, las cenas en mi casa, el Delicias, los vídeos de las tesis, las listas y los emails con los nombres en negrita. Dejo IFAE con millones de recuerdos a mis espaldas. Por todos ellos, muchísimas gracias.

I would also like to thank the people that form MAGIC and make it run: Martin, Javi and Eduardo (*los maestros*), for your priceless help in La Palma, all the galactic conveners I could work with, Pratik, Ignasi, Taka, Christian, Emma and Roberta (aunque ellas se merecen un párrafo a parte), all the people I shared shifts with and work in the multiple projects, specially Daniela, David Carreto and Galindo, always a pleasure working with you.

Roberta, Emma, gracias de verdad por toda vuestra ayuda y por todas esas discusiones en las que siempre aprendía algo nuevo. Espero de corazón que nuestros proyectos se extiendan muchos años más y poder así seguir aprendiendo y descubriendo cosas nuevas con vosotras.

Afortunadamente el doctorado me ha permitido viajar a lo largo del mundo y trabajar con diversos grupos de los cuales he aprendido mucho. First, I would like to thank Petra Huentemeyer, who opened the doors of the HAWC Collaboration to me. During my stay in Michigan, I could learn about HAWC thanks to her, Michele, Hugo and Hao. Along with all of them, Dustin, Anna and John made those two months a lovely period. Un año después volvía a EE.UU., esta vez para trabajar, no sólo en HAWC, sino también en Fermi. A Pepa y a Daniel Castro quiero agradecerles su ayuda esos tres meses. Pepa, te prometo que algún día te prepararé las croquetas andantes que te debo. Esa etapa en Goddard la disfruté también con Mireia, quien me arrancaba del despacho para tomar un café. Thanks also to Katie and Johnny my wonderful housemates in Greenbelt, who showed me how awesome the Peruvian food is. I also want to thank Jordan Goodman, Andrew Smith, Colas, Dan and Israel, the great HAWC group at the University of Maryland for making me feel welcome and for letting me get deeper in the knowledge of HAWC. Thank you to the MPIK group as well, for your kindness during my stay in Heidelberg.

Gracias a Rose, ma petite, por salvar al panda con un cappuccino o una pantalla de por medio. Da igual lo lejos que estemos, *you & me* siempre sonará en nuestra radio.

Pero sobre todo, gracias a TI. Que empezaste siendo mi compañero de despacho, *mi cadete*, y te acabaste convirtiendo en una parte esencial de mí y de mi vida. Que siempre estás ahí, con tu cariño y tu sonrisa, apoyándome en los mejores y peores momentos. Porque no hay palabras en este mundo para expresar lo que siento por ti y lo feliz que me haces. Por todo lo que hemos vivido juntos y lo que nos queda por vivir, gracias, Rubén. Te quiero con todo mi corazón.

Gracias a mi familia, a mis tíos, mis padrinos y mis primos, por estar siempre ahí. A mi abuelo Paco, que soñó con una nieta de pelo negro y vestido azul, a la que después protegía de camino a casa; a mi abuelo Ramos, por todos los viajes en coche y los partidos de fútbol con botas de goma; a mi abuela Carmen, la persona más alegre que jamás conoceré, por llevarme al Cantón desde pequeñita, por cuidarme todos los años que pudiste. Os echo mucho de menos. Á miña avoa Maruxa, polo seu cariño incondicional. Y Martín, Carlos, Lara, no pierdo la esperanza de que algún día leáis estas palabras y os queráis convertir en *astronautas* como la tía. Finalmente, gracias mamá, papá, por todos los sacrificios que tuvisteis que hacer. No ha sido fácil, pero lo hemos conseguido, *nos hemos doctorado*. Gracias por enseñarme que el trabajo duro y la perseverancia dan siempre sus frutos. Y gracias Brais, por tu alegría, por saber escucharme, por ser el mejor hermano mayor que nadie podría tener. No creo que pueda nunca agradeceros a ninguno de los tres todo lo que habéis hecho para que hoy llegue a escribir estas palabras. Sólo espero que os sintáis orgullosos de lo que hemos conseguido y que sepáis lo mucho que os quiero.

October 2017

Contents

Part II   Microquasars in the Very High-Energy Gamma-Ray Regime

Acronyms

L_{Edd}	Eddington Luminosity
AGILE	Astrorivelatore Gamma ad Imagini Leggero
CALLISTO	CALibrate LIght Signals and Time Offsets
Flute	FLUx vs. Time and Energy
Fluxlc	FLUX and Light Curve
MERPP	MERging and Preprocessing Program
SORCERER	Simple, Outright Raw Calibration; Easy, Reliable Extraction Routines
1FGL	First *Fermi*-LAT catalog
2FGL	Second *Fermi*-LAT catalog
3FGL	Third *Fermi*-LAT catalog
a.s.l.	Above sea level
ADC	Analog-to-digital converter
AGN	Active galactic nuclei
AMC	Active mirror control
AP	After pulse
ARC	Astrophysical Research Consortium
ASIC	Application-Specific Integrated Circuit
ATNF	Australia Telescope National Facility
Az	Azimuth
BH	Black hole
C	Carbon
C.L.	Confidence level
C.U.	Crab units
CC	Central control
CCD	Charge-coupled device
CH	Counting house
CIEMAT	Centro de Investigaciones Energéticas Medioambientales y Tecnológicas
ClusCo	Cluster Control

CMB	Cosmic microwave background
CR	Cosmic ray
CTA	Cherenkov Telescope Array
CV	Cataclysmic variable
CW	Cockcroft–Walton
DAQ	Data Acquisition
DC	Direct current
DRS	Domino Ring Sampler
DRS2	Domino Ring Sampler version 2
DRS4	Domino Ring Sampler version 4
DT	Discriminator threshold
EAS	Extended air shower
EBL	Extragalactic background light
ECC	Embedded camera controller
EM	Electromagnetic
FADC	Flash analog-to-digital converter
Fe	Iron
FFT	Fast Fourier transform
FIR	Far infrared
FoV	Field of view
FPGA	Field-programmable gate array
FR	Full-range energy
FWHM	Full-width half-maximum
GBM	GLAST Burst Monitor
GCN	Gamma-ray Coordinate Network
GPS	Global positioning system
GRB	Gamma-ray burst
GZK	Greisen–Zatsepin–Kuzmin
H	Hydrogen
H.E.S.S.	High Energy Stereoscopic System
HAWC	High Altitude Water Cherenkov
He	Helium
HE	High energy
HGPS	H.E.S.S. Galactic Plane Survey
HID	Hardness–intensity diagram
HJD	Heliocentric Julian Day
HMXB	High-mass X-ray binary
HPD	Hybrid photodetector
HS	Hard state
HV	High voltage
IACT	Imaging Atmospheric Cherenkov Telescope
IC	Inverse Compton
IFAE	Institut de Física d'Altes Energies
IPR	Individual Pixel Rate
IR	Infrared

IRF	Instrument response function
IS	Intermediate state
ISM	Interstellar medium
JD	Julian Day
KoP	Key Observation Program
L0	Level 0
L1	Level 1
L3	Level 3
LAT	Large Area Telescope
LE	Low energy
LED	Light-emitting diode
LGS-AO	Laser guide star adaptive optics
LIDAR	Light detection and ranging
LMXB	Low-mass X-ray binary
LST	Large-size telescope
LUT	Look-up table
LVDS	Low-voltage differential signaling
MAGIC	Major Atmospheric Gamma-ray Imaging Cherenkov
MARS	MAGIC Analysis and Reconstruction Software
MC	Monte Carlo
MHD	Magnetohydrodynamic
MJD	Modified Julian Day
MOLA	MAGIC OnLine Analysis
MoU	Memorandum of understanding
MST	Medium-size telescope
Ne	Neon
NIR	Near infrared
NKG	Nishimura–Kamata–Greisen
NN	Next neighbor
NS	Neutron star
NSB	Night sky background
O	Oxygen
OfWP	Off from Wobble Partner
OVRO	Owens Valley Radio Observatory
PACTA	Pre-Amplifier for the Cherenkov Telescope Array
PCB	Printed circuit board
PDB	Power distribution box
PDF	Probability density function
phe	Photoelectron
PMT	Photomultiplier tube
PSF	Point spread function
PSU	Power supply unit
PWN	Pulsar wind nebula
QC	Quality control
QE	Quantum efficiency

RF	Random forest
RMS	Root mean square
RoI	Region of interest
RPP	Rotation-powered pulsar
SDM	Standard disk model
SED	Spectral energy distribution
Si	Silicon
SN	Supernova
SNR	Supernova remnant
SPD	Surge protection device
SPE	Single photoelectron
SS	Soft state
SSC	Synchrotron self-Compton
SST	Small-size telescope
TIB	Trigger interface board
TPU	Trigger processing unit
TS	Test statistics
UB	Universitat de Barcelona
UCL	University College London
UHE	Ultra-high energy
UL	Upper limit
UPS	Uninterruptible power supply
UV	Ultraviolet
VERITAS	Very Energetic Radiation Imaging Telescope Array System
VHE	Very high energy
WCD	Water Cherenkov detector
WD	White dwarf
WR	Wolf–Rayet
Zd	Zenith distance

List of Figures

List of Tables

Part I
Introduction to the Non-thermal Universe

Fig. I.1 Multiwavelength view of the Milky Way. Credit: NASA/Goddard Space Flight Center

Chapter 1
Cosmic Rays and Gamma-Ray Astrophysics

The history of gamma-ray astronomy, or the non-thermal astrophysics in general, is quite recent. It goes back only to the 20th century, when the physicist Victor Hess discovered what was thought to be a new kind of radiation that would forever change our conception of the Universe. In 1912, with the aid of balloon flights, Hess measured that the density of ionized particles increased with altitude (Hess 1912). Some years later, given that its origin was attributed to sources that lay beyond the Earth's atmosphere, Robert Millikan would name this radiation Cosmic Rays (CRs) (Millikan and Cameron 1926). CRs were not strictly a new type of radiation: they are mostly composed in a 99% by protons and Helium (He) nuclei and, in a minor fraction, by heavier nuclei, electrons, positrons and neutrinos. However, CRs present the highest energies observed so far, reaching $\sim 10^{21}$ eV. This highly energetic radiation cannot be thermal in origin and therefore, different processes are needed to explain CR production, related to the most extreme phenomena in the Universe.

CRs offer us a new window from which to observe the Universe, but they give no clue on the astrophysical sources that produce them. Because CRs are composed of charged particles, they are deflected by the randomly oriented magnetic fields they cross in their travel to Earth, erasing any trace of their origin. This is at least true for CRs with energies below 10^{20} eV: above this threshold, particles can travel several Mpc with negligible deviation due to the magnetic fields. Consequently, CRs with energies below EeV are detected randomly distributed when they reach us. By observing them, we can obtain information of their spectrum and composition, but in order to deduce their origin, we need to study neutral particles related to them. To do so, we look for the non-thermal products of the CR acceleration, such as neutrinos and gamma rays. Their neutral charge allows us to trace back their origin.

This thesis is focused on VHE gamma-ray astrophysics. The relationship between CRs and gamma rays, along with the processes and sources from which these VHE gamma rays originate are presented in this section.

© Springer Nature Switzerland AG 2018
A. Fernández Barral, *Extreme Particle Acceleration in Microquasar Jets and Pulsar Wind Nebulae with the MAGIC Telescopes*, Springer Theses, https://doi.org/10.1007/978-3-319-97538-2_1

1.1 Cosmic Rays

The term CRs is nowadays used to define the energetic particles arriving from outside our atmosphere. As mentioned before, they are composed mainly by protons and He nuclei, while only 1% is formed by electrons, positrons, neutrons, neutrinos and heavier nuclei. Gamma rays are also considered part of the CRs. They can be measured directly with balloon experiments or satellites. Normally these techniques explore CRs up to 10^{14} eV. Higher energies can be studied by indirect observations in ground based experiments. These detectors observe the secondary particles produced in the Extended Air Showers (EASs), i.e. cascades of particles originated by the interaction of primary CRs with the nuclei of the Earth's atmosphere. These EASs, first detected by Pierre Auger in 1938, can extend hundreds of meters on the ground. The study of CRs led to important discoveries, such as the existence of the positron (e^+), muons and pions. However, many questions lack for an answer yet, as e.g. *which are the sources responsible for the CRs production?*

1.1.1 Spectrum

The CR spectrum extends from 10^8 to 10^{21} eV, approximately (see Fig. 1.1). Particles with energy below ~ 1 GeV have solar origin, as the solar wind blocks particles at those energies arriving from outside from the solar system. The rest of the spectrum is well-defined by a simple power law, $dN/dE \propto E^{-\Gamma}$, with 3 different photon indices Γ. The first part, from ~ 100 MeV up to ~ 5 PeV, presents a photon index of $\Gamma \sim 2.7$. Its upper limit is known as *knee*, which is charge dependent: particles with higher charge will extend the *knee* to higher energies. The second region covers the range from the *knee* up to ~ 3 EeV, the so-called *ankle*, in which the spectrum follows a power law with $\Gamma \sim 3$. Finally, beyond ~ 3 EeV, the spectrum hardens again (with $\Gamma \sim 2.6$) up to ~ 30 EeV. The different slopes are thought to be related with the origin of the CRs. Particles with energies covering the first range up to the *knee* are believed to be accelerated inside our Galaxy, whilst particles above the *ankle* seem to have extragalactic origin. Between the *knee* and the *ankle* the origin is not that clear. Several proposals were made to explain the steepening of the spectrum at the *knee*: changes on the acceleration mechanism, as e.g. two-step acceleration in the Supernova Remnants (SNRs), first at the front shock and re-acceleration in the inner pulsar-driven remnant (Bell 1991); leakage of CRs out of the Galaxy by diffuse propagation (Ptuskin et al. 1993) or even a cutoff of light elements (Antoni et al. 2005).

The CR spectrum is affected at the edge by the so-called Greisen-Zatsepin-Kuzmin (GZK) cutoff (Fig. 1.2). This cutoff is produced by the interaction of the CRs with energies $\gtrsim 10^{20}$ eV with the Cosmic Microwave Background (CMB):

$$p + \gamma_{CMB} \rightarrow p + \pi^0 \qquad (1.1)$$

Fig. 1.1 CR spectrum obtained with data from different experiments. Credit: Hanlon (2010)

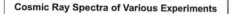

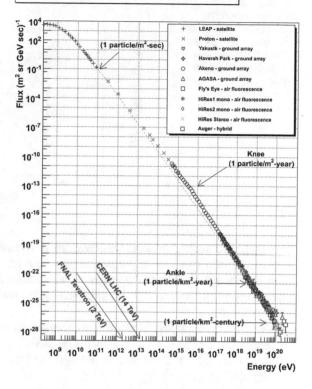

$$p + \gamma_{CMB} \rightarrow n + \pi^0 \tag{1.2}$$

This interaction also limits the maximum distance that the CRs with energies greater than 10^{20} eV can travel to $\sim$50 Mpc.

1.1.2 Cosmic Ray Acceleration

CRs, as charged particles, can be accelerated within magnetic and electric fields. Nevertheless, it is thought that the bulk of CRs are accelerated through diffuse shock acceleration mechanisms. These processes can be split into two main mechanisms, proposed by Fermi (1949):

- **First order Fermi acceleration**: The acceleration takes place in a plasma that presents shock waves (blobs of material moving at supersonic velocities) and magnetic field inhomogeneities. The particles get accelerated every time they cross the shock wave. The energy gained in each reflection is proportional to the relative velocity between the shock and the particle, $\langle \Delta E / E \rangle \propto v_{rel}/c$, and therefore the

Fig. 1.2 CR flux
measurements by Auger and
Telescope Array (TA)
Collaboration, where a cutoff
at energies $\sim 10^{21}$ eV is
evidenced. Taken from
Kampert and Tinyakov
(2014)

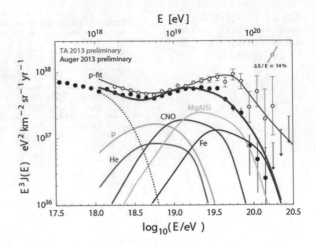

larger the difference, the larger the energy gain. The number of times that a particle crosses the shock is directly proportional to the magnetic field strength. Thus, the crossing frequency is higher as larger the magnetic field strength is. This efficient mechanism is thought to be the responsible of the particle acceleration up to TeV and PeV ranges.

- *Second order* **Fermi acceleration**: This acceleration happens within moving magnetized clouds. The energy gained by the particles in each interaction with the magnetized material is proportional to the square of the speed of the moving cloud, $\langle \Delta E / E \rangle \propto (v_{cloud}/c)^2$.

If particles escape from the acceleration region, they will not be able to gain more energy. Thus, the maximum energy that the accelerated CRs can reach is limited by the radius of the circular motion they describe under the presence of an uniform magnetic field, the so-called Larmor radius or gyroradius. This gyroradius cannot exceed the size of the acceleration region, otherwise the particle would not be confined on this region anymore. This geometrical constraint is known as *Hillas criterion*. The maximum energy, E_{max}, can be expressed as follows:

$$E_{max} \simeq 10^{18} \text{eV} \; q \left(\frac{R}{\text{kpc}} \right) \left(\frac{B}{\mu G} \right) \tag{1.3}$$

where q is the charge of the particle, R and B are the radius and magnetic field of the acceleration region, respectively.

Figure 1.3, the so-called *Hillas plot*, shows the relation between the magnetic field strength and the radius of the acceleration region. The diagonal lines confine the allowed region for acceleration of different particles with certain energy.

Fig. 1.3 Hillas plot that depict the possible CR sources as a function of their magnetic field strength and size. The lines indicate the allowed acceleration region for different particles at a maximum energy (solid red line for protons with $E_{max} = 1$ ZeV $= 10^{21}$ eV, dashed red line for protons with $E_{max} = 100$ EeV $= 10^{20}$ eV, and solid green line for Fe nuclei with $E_{max} = 100$ EeV $= 10^{20}$ eV). Objects below each line cannot accelerate those particles up to the indicated energies Hillas (1984)

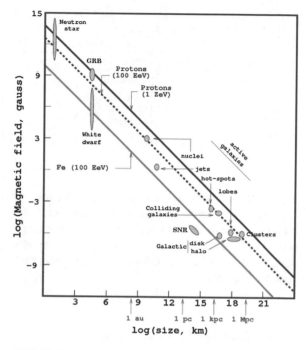

Table 1.1 Classification of the energy domain of the gamma-ray astrophysics

Domain	Abbreviation	Energy range
Low energy	LE	1 MeV–30 MeV
High energy	HE	30 MeV–50 GeV
Very-high energy	VHE	50 GeV–100 TeV
Ultra-high energy	UHE	100 TeV–100 PeV
Extremely-high energy	EHE	>100 PeV

1.2 Gamma-Ray Astrophysics

Gamma rays, photons with the highest energies, arc produced in the acceleration of CRs or their interaction with the environment. The gamma-ray spectrum extend beyond ∼1 MeV, minimum energy of gamma rays produced in electron/positron annihilation ($2 \times m_e = 2 \times 0.511$ MeV, where m_e is the mass of the electron). The different gamma-ray domains are listed in Table 1.1.

Up to now, only gamma rays belonging to the first three energy domains quoted in Table 1.1 were detected. However, given that the CR spectrum extends up to ZeV energies, gamma rays with energies greater than 100 TeV are expected.

Given the high energies that the gamma rays achieve, thermal mechanism cannot be responsible for their production. We need to evoke non-thermal processes to

Table 1.2 Classification of gamma-ray production and absorption mechanisms according to the interaction targets of CRs and gamma rays

Interaction with matter		Interaction with magnetic fields		Interaction with photon fields	
Production	Absorption	Production	Absorption	Production	Absorption
Bremsstrahlung	–	Synchrotron	–	IC	Pair production
e^-/e^+ annihilation π^0 decay		Curvature radiation		SSC	

explain their origin. The interaction of CRs and gamma rays with their environment can be encompassed into three categories: interaction with matter, interaction with magnetic fields and interaction with photon fields. The gamma-ray production processes (Sect. 1.2.1), as well as the gamma-ray absorption (Sect. 1.2.2), will then depend on these targets. Table 1.2 lists the production and absorption mechanisms according to them.

1.2.1 Gamma-Ray Production

The main non-thermal processes that give rise to gamma rays are quoted in Table 1.2 and represented in Fig. 1.4. In the following, I will summarize the main features of each mechanism. For a comprehensive review of the topic, the reader is referred to Aharonian (2004).

1.2.1.1 Bremsstrahlung

Gamma-ray emission is produced when a charged particle is accelerated in the electric field of a nucleus (Fig. 1.4a). This mechanism is more efficient when the charge particle is an electron. Bremsstrahlung becomes dominant against the ionization above the so-called *critical energy*, at which the energy loss by both mechanisms is equal. The mean bremsstrahlung energy loss as a function of the distance traveled by the electron in the medium is given by:

$$-\frac{dE}{dx}\bigg|_{Brems} = \frac{1}{\chi_0}E \tag{1.4}$$

where χ_0 is the *radiation length*, which represents the average distance over which the electron loses all but $1/e$ of its energy due to bremsstrahlung. The *radiation length* depends on the material: low/high density materials will present larger/smaller

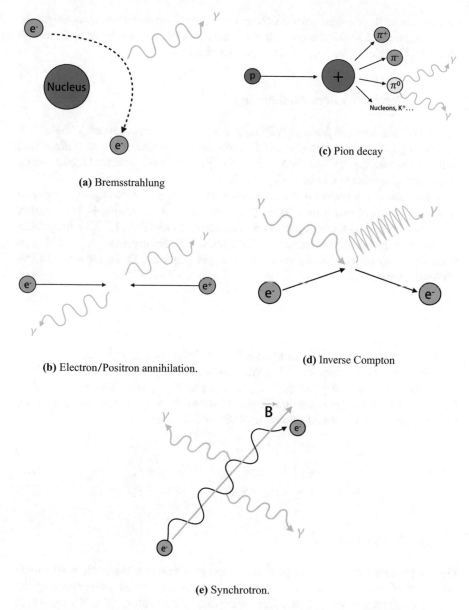

(a) Bremsstrahlung

(c) Pion decay

(b) Electron/Positron annihilation.

(d) Inverse Compton

(e) Synchrotron.

Fig. 1.4 VHE gamma-ray production processes. Modified plots from López-Coto (2015)

radiation length. This implies that the charged particles penetrate longer without losing energy in mediums with lower density.

Taken into account that the energy loss is directly proportional to the energy of the charged particles, the bremsstrahlung gamma-ray spectrum follows the same distribution as the charged particle. Bremsstrahlung is an important mechanism for the production of MeV photons. Gamma rays at VHE can be generated through bremsstrahlung of Ultra-High-Energy (UHE) charges.

1.2.1.2 Electron/Positron Annihilation

When a HE electron and positron collide, two gamma rays are produced ($e^- + e^+ \to 2\gamma$, see Fig. 1.4b). Taken into account the electron and positron mass, this interaction will give rise to a spectral line peak at 511 keV. Nevertheless, given the kinetic energy of the particles this peak is normally broad.

The origin of the positrons is not clear yet, although the simplest possible source is the direct decay of positive pions (Eq. 1.6). Another possibility is the creation of electron/positrons through photon-photon annihilation (Sect. 1.2.2.1), happening e.g. inside of low-mass microquasars (see Sect. 6.1.1). Finally, positrons could arise from the decay of long-lived radioactive isotopes produced in the Supernova (SN) nucleosynthesis.

1.2.1.3 Pion Decay

Pion decay is the dominant hadronic mechanism for gamma-ray production. The interaction takes place between a CR proton with ambient protons or nuclei, resulting on emission of charged pions ($\pi^\pm$) or neutral pions (π^0) with the same probability (Fig. 1.4c). Whilst positive and negative pions decay into muons and neutrinos, neutral pions decay into two gamma rays 99% of the time:

$$\pi^0 \to \gamma + \gamma \tag{1.5}$$

$$\pi^+ \to \mu^+ + \nu_\mu \text{ and } \mu^+ \to e^+ + \bar{\nu}_\mu + \nu_e \tag{1.6}$$

$$\pi^- \to \mu^- + \bar{\nu}_\mu \text{ and } \mu^- \to e^- + \nu_\mu + \bar{\nu}_e \tag{1.7}$$

1.2.1.4 Synchrotron

Due to Lorentz force, charged particles in the presence of a magnetic field follow spiral traces around the magnetic field lines, being accelerated and consequently, emitting radiation (see Fig. 1.4e). This mechanism is more efficient in electrons than in protons. The synchrotron emission does not normally reach VHE, but it can work as target for other processes.

Curvature on the magnetic field lines can lead to **curvature radiation**. This process is similar to synchrotron, but the particles follow the curved magnetic field lines instead of describing spiral paths around them.

1.2.1.5 Inverse Compton

This seems to be the most effective mechanism for production of VHE gamma rays. In this mechanism, relativistic electrons transfer most of their energy to the low-energy photons with which interact, turning them into VHE gamma rays (Fig. 1.4d). According to the energy of both, electrons and target photons (E_e and E_γ, respectively), one can distinguish two regimes:

- **Thomson regime**: This regimes happens when $E_\gamma E_e << m_e^2 c^4$, leading to a constant cross-section of $\sigma_T = \frac{8}{3}\pi r_e^2$, where r_e is the electron radius. For a power-law distribution of electrons with photon index Γ_e, the up-scattered gamma-ray spectrum follows a power-law function as well with index $\Gamma = (\Gamma_e + 1/2)$ (Ginzburg and Syrovatskii 1964). The energy loss in this regime is proportional to E_e^2 ($dE_e/dt = \frac{4}{3}\sigma_T c E_\gamma n_\gamma E_e^2$, where n_γ is the density of initial photons).
- **Klein-Nishima regime**: This regime is considered when $E_\gamma E_e \approx m_e^2 c^4$, for which the cross-section can be defined as:

$$\sigma_{KN} = 2\pi r_e^2 \left\{ \frac{1+\varepsilon}{\varepsilon} \left[\frac{2+2\varepsilon}{1+2\varepsilon} - \frac{ln(1+2\varepsilon)}{\varepsilon} \right] + \frac{ln(1+2\varepsilon)}{2\varepsilon} - \frac{1+3\varepsilon}{(1+3\varepsilon)^2} \right\} \quad (1.8)$$

where $\varepsilon = E_\gamma/m_e c^2$. In cases where $E_\gamma E_e >> m_e^2 c^4$, one can follow the Klein-Nishima approximation where $\sigma_{KNapprox} = \frac{1}{\varepsilon}\pi r_e^2 \left[ln(2\varepsilon) + \frac{1}{2} \right]$. In the Klein-Nishima regime, the resulting gamma-ray spectrum, given an electron spectrum again well-fitted by a power-law with Γ_e, is considerably steeper, with a photon index of $\Gamma = \Gamma_e + 1$. The energy loss is here independent from the electron energy, but proportional to the density of photons.

When the seed photons on the Inverse Compton (IC) scattering are synchrotron gamma rays emitted by the same electron population, then the process is called **Synchrotron Self Compton (SSC)**. This mechanism allows synchrotron radiation to reach the VHE regime.

1.2.2 Gamma-Ray Absorption

1.2.2.1 Pair Production

Gamma rays also suffer from annihilation, which decreases their flux and make more difficult their detection in some scenarios. The main gamma-ray absorption mechanism is the so-called pair production, in which a HE photon interacts with a lower energy one to give rise to an electron/positron pair. It is therefore the inverse process from the pair annihilation (Sect. 1.2.1.2). There are two types of pair production:

- **Classical pair production**: This process is the responsible for the cascades produced in our atmosphere (see Chap. 2). It occurs when a HE photon interacts with a virtual photon of a nucleus' electric field:

$$\gamma(\gamma) \rightarrow e^- + e^+ \tag{1.9}$$

- **Photon-photon annihilation**: Interaction between HE photons with lower-energy photons from the ambient gas. This process is the responsible from the attenuation of extragalactic VHE by the Extragalactic Background Light (EBL) and plays an important role in the absorption of galactic VHE gamma rays, as e.g. inside the binary systems:

$$\gamma\gamma \rightarrow e^- + e^+ \tag{1.10}$$

The cross-section of the pair production presents a peak at:

$$E_{\gamma 1} E_{\gamma 2}(1 - cos\theta) \sim 2(m_e c^2)^2 \tag{1.11}$$

where $E_{\gamma 1}$ and $E_{\gamma 2}$ are the energy of the seed photons, θ is the collision angle between them and $m_e = 0.511$ MeV is the mass of the electron. This implies that the highest cross-section and hence, the highest probability of absorption for a gamma ray of ~ 100 GeV takes place with a photon around the Infrared (IR) and Ultraviolet (UV) band. This is the reason why VHE photons created in the microquasars jets might be strongly affected by the stellar wind, whose emission peaks at those lower energies. On the other hand, HE photons from these sources are more affected at the base of the jet where the soft X-ray population is higher (see Sect. 3.2.4).

1.2.3 Gamma-Ray Sources

The processes shown in Sect. 1.2.1 can occur in different astrophysical objects. Here I provide a brief description of the already established gamma-ray sources, divided according to their galactic or extragalactic nature.

1.2.3.1 Galactic Sources

- **SNR**: Leftovers of the SN explosions. Gamma-ray emission is produced by the interaction between CRs, accelerated through the *first order* Fermi mechanism in the shock wave from the SN explosion, and nuclei in the Interstellar Medium (ISM).
- **Pulsars**: Highly magnetized rotating Neutron Star (NS). Particles can be accelerated in very specific regions along the magnetic field, from which gamma-ray emission is produced in narrow beams. Its emission is characterized by a pulsation: since the magnetic fields lines and the rotation axis of the NS are not usually

aligned, the beam emission is only detected when crossing our light of sight. The best studied pulsar is the Crab pulsar, first detected by MAGIC (Aliu et al. 2008) and recently detected up to TeV energies (Ansoldi et al. 2016). Along with Crab, only Vela pulsar has been detected in the gamma-ray band (Brun 2014).

- **PWN**: Magnetized cloud of relativistic particles created when the pulsar wind interacts with the ISM. VHE emission originates through IC scattering of ambient photons by accelerated electrons. An extended description is given in Chap. 7.
- **Gamma-ray binaries**: Binary system composed of a massive star and a compact object, either a BH or a NS, whose peak of luminosity lays in the gamma-ray regime. Their emission is explained by two models: *pulsar wind scenario* and *microquasars*. In the former, the VHE emission arises from the interaction of the compact object wind with the stellar wind. In the microquasar scenario, the compact object accretes material from the companion. The system presents relativistic jets, whose existence depends on the accretion rate, where particles are accelerated and emit gamma rays in the IC interaction with the stellar wind. Five gamma-ray binaries have been detected so far: PSR B1259–63, LS I +61°303, HESS J0632+057, 1FGL J1018.6–58 and LS 5039. None of them belong to the microquasar scenario. Nevertheless, gamma rays from X-ray binaries such as Cygnus X-1 and Cygnus X-3 ranked as microquasars were detected. Detailed information is available at Sect. 3.1.

1.2.3.2 Extragalactic Sources

- **AGN**: Galaxies hosting a super-massive BH in their center, which accretes material from the surrounding. They present two relativistic jets perpendicular to the accretion disk formed around the compact object. As in the case of microquasars, the gamma-ray emission is produced in the jets where charged particles can get accelerated.
- **Starburst galaxies**: Galaxies in which the star formation rate is high. Consequently, the SN explosion rate is large giving rise to a high CR density. The starburst galaxy M82, detected by Very Energetic Radiation Imaging Telescope Array System (VERITAS) (VERITAS Collaboration et al. 2009), hosted the closest Type Ia SN in the last decades, SN 2014J, studied in this thesis (Appendix A).
- **Gamma-Ray Burst (GRB)**: Sudden and short gamma-ray outburst. They correspond to the most energetic gamma-ray flares. Their nature is still unclear, but two scenario have been proposed: collapse of highly rotating very massive stars ($M_\star > 100\ M_\odot$), the so-called hypernovae, or merge of two compact objects.

Table 1.3 Comparison of the different gamma-ray detection techniques

	Satellites	IACTs	Water Cherenkov array
Experiments	*Fermi*-LAT	MAGIC, H.E.S.S. VERITAS, FACT	HAWC
Energy range	Few MeV to hundred GeV	Few GeV to few TeV	Hundred GeV to Hundred TeV
Advantages	Excellent γ/h separation	Good γ/h separation	Very large collection area
	Full duty cycle	Excellent angular resolution	Full duty cycle
	Large FoV	Good energy resolution	Very large FoV
Disadvantages	Poor angular resolution (at low E)	Limited duty cycle	Poor γ/h separation (at low E)
	Low collection area	Reduced FoV	Poor angular and energy resolution

1.2.4 Detection Techniques

In order to detect the broadband gamma-ray spectrum, different detection techniques have been applied. Focusing on the HE and VHE regime, one can divide these techniques into three types: satellites, which perform direct observations of the gamma rays above the Earth's atmosphere, Imaging Atmospheric Cherenkov Telescopes (IACTs) and Water Cherenkov arrays, which perform indirect observations at the ground level. The main characteristic of these techniques and the current facilities using them are summarized in Table 1.3.

- **Satellites**: Due to the reduced collection area that detectors on-board satellites have ($\sim$1 m^2), this technique provides information from MeV to a few hundred GeV. Pair production is the dominant mechanism for the detection of gamma rays above $\sim$30 MeV. These detectors present wide Field of View (FoV) that observe during a $\sim$100% duty cycle. They provide excellent γ/hadron separation, but with a poor angular resolution at low energies ($>$0.5$°$ below $\sim$ GeV). On the other side, their energy resolution is very good, with small systematic errors.
- **IACTs**: This technique is based on the detection of Cherenkov light produced in the electromagnetic cascades originated by the interaction of a primary gamma ray with the nuclei of our atmosphere. An array of this type of telescopes allows to increase the collection area up to several km^2 and therefore, the energy range increases considerably compared to the one covered by satellites. In turn, this technique provides very good γ/hadron separation and excellent angular resolution. Nevertheless, the high background light do not permit to observe during daytime.
- **Water Cherenkov arrays**: Particles in the EASs produced by VHE gamma rays (at energies greater than $\sim$100 GeV) can reach the ground. Gamma rays can be

studied by the indirect observation of the Cherenkov light produced when these particles cross water tanks. Their very large collection area allows to detect multi-TeV gamma-rays. However, they present the worst angular resolution among all detection techniques in the GeV regime, around $\sim 1°$. Their energy resolution is as well worse than the one achieved by IACTs. Nonetheless, they have full duty cycles, since daytime observations are possible, and large FoV.

References

Aharonian FA (2004) Very high energy cosmic gamma radiation: a crucial window on the extreme Universe. World Scientific Publishing Co

Aliu A, Antonelli A, Backes B, Barrio B, Bastieri B, Bednarek B, Bernardini B, Biland B, Bonnoli B, Bosch-Ramon B, Britvitch C, Carmona C, Commichau C, Cortina C, Covino C, Dazzi DA, de Cea Del Pozo E, de los Reyes R, de Lotto B, de Maria M, de Sabata F, Delgado Mendez C, Dominguez A, Dorner D, Doro M, Elsässer, Errando M, Fagiolini M, Ferenc D, Fernandez E, Firpo R, Fonseca MV, Font L, Galante N, Garcia Lopez RJ, Garczarczyk M, Gaug M, Goebel F, Hadasch, Hayashida M, Herrero A, Höhne D, Hose J, Hsu CC, Huber S, Jogler T, Kranich D, La Barbera A, Laille A, Leonardo E, Lindfors E, Lombardi S, Longo F, Lopez E, Lorenz E, Majumdar P, Maneva G, Mankuzhiyil N, Mannheim K, Maraschi L, Mariotti M, Martinez M, Mazin D, Meucci M, Meyer M, Miranda J, Mirzoyan R, Moles M, Moralejo A, Nieto D, Nilsson K, Ninkovic J, Otte N, Oya I, Paoletti R, Paredes JM, Pasanen M, Pascoli D, Pauss F, Pegna RG, Perez-Torres MA, Persic M, Peruzzo L, Piccioli A, Prada F, Prandini E, Puchades N, Raymers A, Rhode W, Ribó M, Rico J, Rissi M, Robert A, Rügamer S, Saggion A, Saito TY, Salvati M, Sanchez-Conde M, Sartori P, Satalecka K, Scalzotto V, Scapin V, Schweizer T, Shayduk M, Shinozaki K, Shore, Sidro N, Sierpowska-Bartosik A, Sillanpää A, Sobczynska D, Spanier F, Stamerra A, Stark LS, Takalo L, Tavecchio F, Temnikov P, Tescaro D, Teshima M, Tluczykont M, Torres DF, Turini N, Vankov H, Venturini A, Vitale V, Wagner RM, Wittek W, Zabalza V, Zandanel F, Zanin R, Zapatero J, de Jager, de Ona Wilhelmi, MAGIC Collaboration, Aliu E et al (2008) Science 322:1221

Ansoldi S, Antonelli LA, Antoranz P, Babic A, Bangale P, Barres de Almeida U, Barrio JA, Becerra González J, Bednarek W, Bernardini E, Biasuzzi B, Biland A, Blanch O, Bonnefoy S, Bonnoli G, Borracci F, Bretz T, Carmona E, Carosi A, Colin P, Colombo E, Contreras JL, Cortina J, Covino S, Da Vela P, Dazzi F, De Angelis A, De Caneva G, De Lotto B, de Oña Wilhelmi E, Delgado Mendez C, Di Pierro, Dominis Prester D, Dorner D, Doro M, Einecke S, Eisenacher D, Glawion, Elsaesser D, Fernández-Barral, Fidalgo, Fonseca MV, Font L, Frantzen K, Fruck C, Galindo D, García López RJ, Garczarczyk M, Garrido Terrats D, Gaug M, Godinović N, González Muñoz A, Gozzini SR, Hanabata Y, Hayashida M, Herrera J, Hirotani D, Hose J, Hrupec D, Hughes G, Idec W, Kellermann H, Knoetig ML, Kodani K, Konno Y, Krause J, Kubo H, Kushida J, La Barbera A, Lelas D, Lewandowska N, Lindfors E, Lombardi S, Longo, López M, López-Coto R, López-Oramas R, Lorenz E, Makariev M, Mallot K, Maneva G, Mannheim N, Maraschi L, Marcote B, Mariotti M, Martínez M, Mazin D, Menzel U, Miranda JM, Mirzoyan R, Moralejo A, Munar-Adrover P, Nakajima D, Neustroev, Niedzwiecki, Nevas Rosillo, Nilsson K, Nishijima K, Noda K, Orito R, Overkemping A, Paiano S, Palatiello M, Paneque D, Paoletti R, Paredes JM, Paredes-Fortuny X, Persic M, Poutanen, Prada Moroni PG, Prandini E, Puljak I, Reinthal R, Rhode W, Ribó M, Rico J, Rodriguez Garcia J, Saito T, Saito K, Satalecka K, Scalzotto V, Scapin V, Schultz C, Schweizer T, Shore SN, Sillanpää A, Sitarek J, Snidaric I, Sobczynska D, Stamerra A, Steinbring T, Strzys M, Takalo L, Takami H, Tavecchio F, Temnikov P, Terzić T, Tescaro D, Teshima M, Thaele J, Torres DF, Toyama T, Treves A, Ward, Will, Zanin Ansoldi S et al (2016) 585:A133

Antoni A, Badea B, Bercuci B, Bozdog B, Chilingarian D, Doll E, Engler F, Gils G, Haungs H, Hörandel K, Klages M, Mathes M, Milke M, Obenland O, Ostapchenko P, Rebel R, Risse R, Schatz S, Scholz T, Ulrich H, van Buren J, Vardanyan A, Weindl A, Wochele J, Zabierowski J, Antoni T et al (2005) Astropart Phys 24:1
Bell AR (1991) International cosmic ray conference, vol 2, p 420
Brun P (2014) Very high energy phenomena in the universe
Fermi E (1949) Phys Rev 75:1169
Ginzburg VL et al (1964) The origin of cosmic rays
Hanlon WF (2010) http://www.physics.utah.edu/whanlon/spectrum.html
Hess VF (1912) Z Phys 13:1084
Hillas AM (1984) 22:425
Kampert K-H et al (2014) C R Phys 15:318
López-Coto R (2015) Very-high-energy γ-ray observations of pulsar wind nebulae and catacysmic variable stars with MAGIC and development of trigger systems for IACTs. PhD thesis
Millikan RA et al (1926) Phys Rev 28:851
Ptuskin VS et al (1993) 268:726
VERITAS Collaboration, Acciari VA, Aliu E, Arlen T, Aune T, Bautista M, Beilicke M, Benbow W, Boltuch M, Bradbury SM, Buckley JH, Bugaev V, Byrum K, Cannon A, Celik O, Cesarini L, Chow YC, Ciupik L, Cogan P, Colin P, Cui W, Dickherber R, Duke C, Fegan SJ, Finley JP, Finnegan G, Fortin P, Fortson L, Furniss A, Galante N, Gall D, Gibbs K, Gillanders H, Godambe S, Grube J, Guenette R, Gyuk G, Hanna D, Holder J, Horan D, Hui CM, Humensky TB, Imran A, Kaaret P, Karlsson N, Kertzman M, Kieda D, Kildea J, Konopelko A, Krawczynski H, Krennrich F, Lang MJ, Lebohec S, Maier G, McArthur S, McCann A, McCutcheon M, Millis J, Moriarty P, Mukherjee R, Nagai T, Ong RA, Otte AN, Pandel D, Perkins JS, Pizlo F, Pohl M, Quinn J, Ragan K, Reyes LC, Reynolds PT, Roache E, Rose HJ, Schroedter M, Sembroski GH, Smith AW, Steele D, Swordy SP, Theiling M, Thibadeau S, Varlotta A, Vassiliev VV, Vincent S, Wagner RG, Wakely SP, Ward JE, Weekes TC, Weinstein A, Weisgarber T, Williams DA, Wissel S, Wood M, Zitzer B, VERITAS Collaboration et al (2009) 462:770

Chapter 2
Gamma-Ray Telescopes

In this chapter, I will describe in detail both hardware of the Major Atmospheric Gamma-ray Imaging Cherenkov (MAGIC) telescopes and the software used in the analysis of the data. Moreover, I will give an overview of the planned CTA, future generation of Cherenkov telescopes. Although this thesis focuses on the work performed with these two IACT arrays, other gamma-ray detection techniques were used in the analysis and/or discussion of several sources included here. Therefore, I will also describe briefly in this chapter the *Fermi*-LAT detector and the wide FoV observatory High Altitude Water Cherenkov (HAWC).

Before delving into the MAGIC and CTA characteristics, I will start giving a general view of the imaging atmospheric Cherenkov technique.

2.1 Cherenkov Light

Gamma rays correspond to the highest radiation in the Electromagnetic (EM) spectrum, which covers 20 energy decades between radio and the TeV regime. Some energies can travel across the atmosphere reaching the ground, but the interaction of the gamma rays with the molecules in the atmosphere prevents the most energetic radiation from penetrating and reaching us. In order to detect these high-energy photons, detectors on-board satellites can be used. Nevertheless, due to weight limitations, they can only support detectors with small collection area and therefore, they cannot provide results for energies above hundred GeV, regime in which the photon flux is already low.

For energies $\gtrsim$50 GeV, IACTs (with larger sensitive area) dominate the study of gamma rays. The technique is based on indirect detection. When a VHE gamma ray or CR interacts with the atmospheric nuclei, a particle cascade is initiated, the so-called EAS. If the resulting charged particles of this interaction travel faster than

© Springer Nature Switzerland AG 2018
A. Fernández Barral, *Extreme Particle Acceleration in Microquasar Jets and Pulsar Wind Nebulae with the MAGIC Telescopes*, Springer Theses, https://doi.org/10.1007/978-3-319-97538-2_2

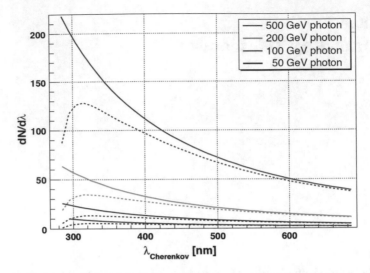

Fig. 2.1 Spectra of Cherenkov radiation produced by vertical EAS initiated by gamma rays at different energies. The solid lines corresponds to the unabsorbed spectra at 10 km altitude, while the dashed line are the observed spectra attenuated by Rayleigh and Mie scattering (see Sect. 2.1)

the speed of light in the atmosphere, Cherenkov light is emitted, whose wavelength ranges between 300 and 500 nm. The existence of this type of light was proposed by the Soviet physicist Pavel Alekseyevich Cherenkov (Cherenkov 1934) who, along with Ilya Frank and Igor Tamm, received the Nobel prize in 1958 for the discovery and interpretation of the Cherenkov effect.

The Cherenkov radiation peaks at $\sim$320 nm, i.e. in the UV band (see Fig. 2.1). However, the emitted and observed radiation spectra differ due to the transmission losses in the atmosphere. The main sources of this attenuation are:

- **Rayleigh scattering**: Scattering off air molecules, with a wavelength dependency of λ^{-4}. It affects mostly UV radiation.
- **Mie scattering**: Scattering off aerosols, dust and droplets water. It does not show any strong wavelength dependency.
- **Ozone molecules**: These molecules are responsible for the strong absorption of hard UV photons (<300 nm).
- **H_2O and CO_2 molecules**: They produce absorption in the IR band.

There is also a dependency on the zenith angle of EAS: the higher the angle, the higher the attenuation. This is due to the fact that at high zenith angle, the cascades develop in the highest layers of the atmosphere and hence, particles need to travel a larger path. Consequently, the probability of suffering absorption from some of the above mentioned processes increases. Only EASs initiated by particles at the highest energies are significantly detected by the telescopes at high Zenith distance (Zd) range. Thus, at larger zenith ranges the peak of the Cherenkov radiation spectrum

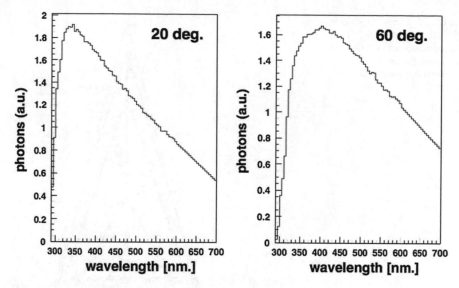

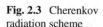

Fig. 2.2 Cherenkov spectra at different zenith angles

Fig. 2.3 Cherenkov
radiation scheme

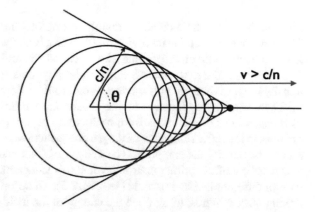

shifts to larger wavelengths. Figure 2.2 shows how the density of Cherenkov photons, as well as the peak of their spectrum, vary at different zenith angles.

The shape of the Cherenkov radiation around the track of the charged particle is a cone with an aperture angle θ, (the so-called Cherenkov angle; see Fig. 2.3), given by:

$$cos\theta = \frac{c'}{v} = \frac{c}{vn(\lambda)} \tag{2.1}$$

where $c' = c/n$ is the speed of light in the medium and $n(\lambda)$ is the refractive index of the medium, whose value varies with the wavelength (λ) of the Cherenkov light. The mean value of θ in air is $\sim 1°$Àn ultrarelativistic particle propagating vertically

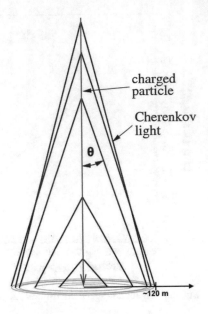

charged particle

Cherenkov light

θ

~120 m

through the atmosphere creates a doughnut ring of Cherenkov light in the ground. The contribution of all the involving particles in a EAS that emit Cherenkov radiation leads to a full circle on the ground, the so-called Cherenkov light pool (Fig. 2.4).

In the case of a vertical EAS initiated by a gamma ray, the Cherenkov photons density is approximately uniform in a circle from the core of the cascade up to $\lesssim$120 m. There is a slightly increase on the density around this distance, which is known as *hump*, whose origin arises from an increase in the opening angle (θ) due to the changes in the refraction index as the particle penetrates the atmosphere. Beyond the *hump*, the density fades rapidly. The density of Cherenkov photons is proportional to the energy of the primary particle when this is a gamma ray, which is not true in case of different incident particle (see Fig. 2.5). Therefore, this relation can be used, among other features, to estimate the energy of the incident gamma ray.

Although we are interested in EASs initiated by gamma rays, cascades induced by hadrons (mainly protons) are much more numerous. Even for strong gamma-ray sources, as it is the case of the Crab Nebula, the ratio between hadron-induced and gamma ray-induced cascades is considerably high, around 1000 hadronic cascades for each electromagnetic shower above hundred of GeV. Therefore, hadronic cascades represent the major source of background in our observations. The better we understand both types of showers, the better we can get rid of the background that embedded our observations.

Fig. 2.5 Cherenkov photon density within a radius of 125 m from the core shower as a function of photons energy for different primary particles. Taken from Wagner (2006)

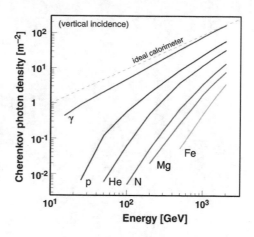

2.2 Types of EAS

The imaging atmospheric Cherenkov technique differentiates between cascades induced by gamma rays and hadrons based on the images that leave on the telescope cameras. These images are different depending on the particle interaction with the atmospheric nuclei and its development along the air, which gives rise to showers with distinct features. In the following sections, I will give an overview of the difference between gamma ray- and hadron-induced EAS.

2.2.1 Electromagnetic Showers

Gamma rays can initiate particle cascades through the pair creation process on air nuclei (see Sect. 1.2.2.1) if their energy is $\gtrsim 20$ MeV. The electrons and positrons, product of this interaction, emit in turn gamma rays via bremsstrahlung (see Sect. 1.2.1.1). The latter takes place until the electrons and positrons reach the so-called *critical energy*, that in air is $E_C = 86$ MeV, below which the ionization energy loss dominates. If photons emitted through bremsstrahlung have enough energy, they undergo pair creation as well, leading to a EM cascade (see Fig. 2.6). The bremsstrahlung *radiation length* for electrons and positrons in air is $\chi_0^e = 37$ g cm^{-2} and the *mean free path* (average distance traveled between collisions) of gamma rays due to pair creation is $\chi_0^\gamma = 7/9\chi_0^e$. Consequently, the particles in an EM shower do not scatter too much from the shower axis, leading to a quite symmetric cascade (see left plots on Fig. 2.7). The cross-section for the interaction of gamma rays with the atmospheric nuclei is weakly dependent on the photon energy, and therefore the height of the collision of the primary gamma ray is similar for different gamma ray energies, being located at $\sim$20–30 km above sea level (a.s.l.). In each step of the shower, the number of particles is doubled, while the particle energy

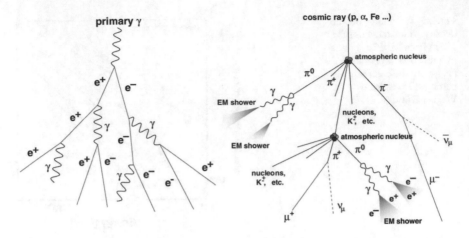

Fig. 2.6 Schemes of an EM (*left*) and hadronic (*right*) showers. Credit: Wagner (2006)

halves, until reaching E_C, moment at which the shower is disrupted and the number of particles reaches its maximum. The altitude at which this condition is fulfilled is called *height of the shower maximum* and it is inversely proportional to the logarithm of the primary gamma-ray energy, $H_{max} \propto 1/ln(E)$.

2.2.2   Hadronic Showers

Hadronic cascades are those produced by the interaction between a cosmic and atmospheric nuclei. Normally, the primary particle of this interaction is a proton which gives rise mostly (~90%) to pions (approximately in the same proportion π^+, π^-, π^0; see Sect. 1.2.1.3). Besides pions, these collisions produce kaons and nuclei (Fig. 2.6). Both hadrons and pions undergo more collisions or decays that generate the shower. The cascade stops when the energy per nucleon is less than ~1 GeV, minimum energy needed for pion production. Different components can be distinguished in the hadronic showers:

- **Hadronic component**: Composed by nuclei and mesons (like pions). Both of them are heavy particles and therefore, the transferred transversal momentum in each collision is high. The following pionic decays take place inside these cascades:

$$\pi^+ \rightarrow \mu^+ \nu_\mu \qquad\qquad (2.2)$$

$$\pi^- \rightarrow \mu^- \bar{\nu}_\mu \qquad\qquad (2.3)$$

$$\pi^0 \rightarrow 2\gamma \qquad\qquad (2.4)$$

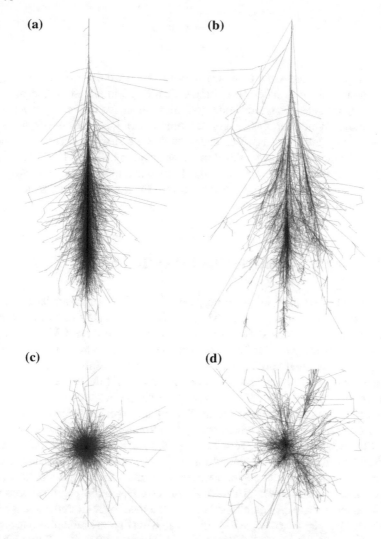

Fig. 2.7 *On the left*: MC simulation of an EM cascade initiated by a 100 GeV gamma ray. *On the right*: MC simulation of a hadronic cascade initiated by a 100 GeV proton. Red lines show the gamma-ray, electron and positron tracks, green lines are used for muons and blue ones for hadrons. The upper plots represent the vertical trajectory, while the lower plots represent the transversal planes

- **EM component**: Composed by secondary photons, electrons and positrons, mostly arriving from the decay of π^0. If these EM subcascades are detected, the distinction between them and a gamma ray-induced shower is almost impossible.
- **Muonic component**: Muons and neutrinos are produced by the decay of charged pions. Neutrinos cannot be detected by Cherenkov telescopes as they cannot produce Cherenkov radiation due to the lack of charge. On the other hand, muons can be detected and also decay into:

$$\mu^+ \rightarrow e^+ \nu_e \bar{\nu}_\mu \tag{2.5}$$

$$\mu^- \rightarrow e^- \bar{\nu}_e \nu_\mu \tag{2.6}$$

Hadronic showers are wider than the EM ones, because the transversal momentum that the kaons and pions received is higher than those in electrons and positrons. Furthermore, this type of cascade undergoes more subshowers, leading to a not only wider but more asymmetric EAS. Figure 2.7 presents MC simulations of gamma ray- (left) and hadron-induced (right) cascades, where the shape difference is evident. On the other hand, EM showers develop faster than the hadronic ones. The former develops in less than 3 ns compared to the 10 ns taken in hadronic cascades. Thus, besides their image feature left on the telescopes, the timing parameter can be used to distinguish these type of showers.

2.3 Imaging Atmospheric Cherenkov Technique

As mentioned before, the imaging atmospheric Cherenkov technique bases its study of gamma rays on the indirect observations of the Cherenkov radiation produced in EAS. In Sect. 2.2 it was shown the main difference between EM and hadronic cascades. The several particles involved in each type lead to different shower shape and timing features that can be used to distinguish them through the Cherenkov telescopes. In Fig. 2.8, we can see the way IACTs work. If the telescopes are inside the Cherenkov light pool, part of the Cherenkov light is reflected in their mirrors and collected in their fast pixelized cameras. The images created in these cameras are projections of the EASs, from which spatial and timing information is obtained.

The Cherenkov radiation is very fast ($\sim$3 ns) for which precise and very efficient detectors are needed. Photomultiplier Tubes (PMT) are commonly used in air Cherenkov telescopes as they have proven to fulfill these requirements. A fast response time is also important to avoid collect undesirable photons, product of the background sources. As mentioned before, the main background sources are the hadronic showers, from which the EM subcascades act like an irreducible background for the gamma-ray observations. Besides the background from the EASs, photons isotropically distributed on the sky can affect the observations as well. This is the so-called Night Sky Background (NSB) that is formed by the stars' light, airglow, polar and zodiacal light and artificial lights. In La Palma, where the MAGIC telescopes are located, this NSB contribution was measured to be $(1.75 \pm 0.4) \times 10^{12}$ photons $m^{-2}sr^{-1}s^{-1}$.

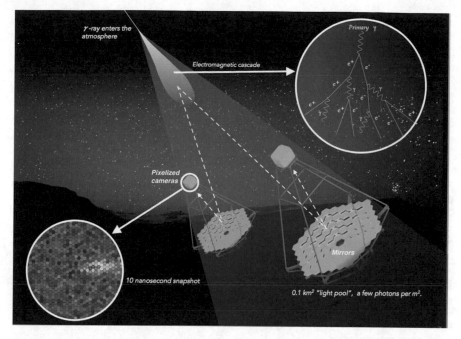

Fig. 2.8 Sketch of the imaging atmospheric Cherenkov technique. The Cherenkov light from the cascade is reflected in the mirrors and collected in the camera. Modified image from www.cta-observatory.org

2.4 MAGIC

MAGIC is a stereoscopic system consisting of two 17 m diameter imaging atmospheric Cherenkov telescopes (IACTs) located in El Roque de los Muchachos in the Canary island of La Palma, Spain (28.8° N, 17.8° W, 2225 m a.s.l.). Until 2009, MAGIC consisted of just one stand-alone IACT with an integral flux sensitivity around 1.6% of the Crab Nebula flux in 50 h of observation (Aliu et al. 2009a). After autumn 2009, the second telescope (MAGIC II) started operation, allowing us to reach in stereo mode an energy threshold as low as 50 GeV at low zenith angles (Aleksić et al. 2012b). In this period the sensitivity improved up to $0.76 \pm 0.03\%$ of the Crab Nebula flux for energies greater than 290 GeV in 50 h of observations. Between summer 2011 and 2012 both telescopes underwent a major upgrade that involved the digital trigger, readout systems and the MAGIC I camera (Aleksić et al. 2016a). After this upgrade, the system achieves, in stereoscopic observational mode, an integral sensitivity of $0.66 \pm 0.03\%$ of the Crab Nebula flux in 50 h above 220 GeV (Aleksić et al. 2016b). The data analyzed in this thesis covered both mono (only MAGIC I) and stereoscopic observations and therefore, both will be described in this section (Fig. 2.9).

Fig. 2.9 Picture of the MAGIC telescopes at El Roque de los Muchachos. Image taken from
https://magic.mpp.mpg.de/

2.4.1   Hardware

In this section, I will give a description of the main components of the MAGIC
telescopes, depicted in Fig. 2.10 and listed below:

- Alt-azimuth frame and drive system.
- Mirrors and reflector.

Fig. 2.10 Picture of the MAGIC telescopes (MAGIC I behind, MAGIC II on the front) with some of
their hardware subsystems highlighted. Background picture taken from https://magic.mpp.mpg.de/,
considering that the subsystems were included by me

- Camera.
- Trigger system.
- Readout system.
- Other subsystems.

The Central Control (CC) software of the telescopes is called SuperArehucas, responsible for all the subsystems. It receives and sends reports, monitoring each hardware subsystem, and provides access to most of the funcionalities of the telescopes through a LabView interface. A complete description of SuperArehucas is available in Zanin (2011).

2.4.1.1 Alt-Azimuth Frame and Drive System

The octogonal telescope structure that supports the 17 m reflector dish is made of light carbon fibre-epoxy tubes which hold together through aluminum knots (see Fig. 2.10). The total weight of the structure, including the reflector and the camera support, is less than 20 tons. This allows a fast repositioning of the telescopes. The camera is held by an aluminum circular tube secured to the main structure by 20 steel cables. The telescopes have an alt-azimuthal mount and therefore, they are moved in two axes, Azimuth (Az) and Zd, when pointing to a source. For any orientation of the telescopes, the deformation of the structure is lower than 3.5 mm (Bretz et al. 2009). Small bending of the structure during the movement of the telescopes are corrected before starting data taking through the Active Mirror Control (AMC) (see Sect. 2.4.1.2).

The Az range covers −90° to 318° and the Zd does so between −70° to 105°. The Az movement on a 20 m diameter circular rail is carried out by two 11 kW motors. The telescopes move on the Zd or elevation axis thanks to one motor of the same power, located behind the dish structure (see Fig. 2.10). For safety reasons, the drive system is automatically stopped to avoid the movement if the fence around the telescopes is open (i.e. somebody is inside the telescopes area).

The position at which the telescopes are pointing is measured by three 14-bit shaft encoders (two in Az, one in Zd). Thus, the telescope position is given with an accuracy of ∼0.02°.Along with the shaft encoders, the pointing of the telescopes is constantly checked during observations with the *Starguider camera*, a Charge-Coupled Device (CCD) camera located in the middle of the dish (Fig. 2.11). It points to the same direction of the telescopes and analyzes the stars in the FoV. This FoV is compared with a star catalog to find any possible mispointing (undesirable shift between the camera center and the real coordinates of the source). The reliability of the *Starguider camera* is measured by the ratio between the number of observed stars and the number of expected stars in a given FoV, which can be disturbed by a low sky visibility (high cloudiness or humidity). The *Starguider camera* observes, along with the sky, part of the PMT camera in which a set of Light-Emitting Diodes (LED) are installed to provide a reference frame (see Fig. 2.12). The star catalog used to correct any offset is the so-called *bending model*. This *bending model* is created

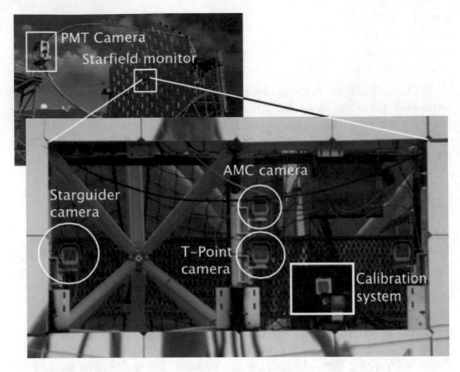

Fig. 2.11 CCD in the middle of the reflector dish. Wagner (2006), Mazin (2007)

comparing cataloged and observed coordinates of ~150 bright stars taken with the *T-Points camera* (another CCD camera located also in the dish, see Fig. 2.11) at the end of each night (~5 stars/night). It is updated every observational period (period between two consecutive full moon breaks). Thus, making use of this model and the shaft encoders, a pointing precision of 0.01° is achieved.

2.4.1.2 Mirrors and Reflector

The 17 m reflectors used in MAGIC are parabolic and therefore, isochronous. This is important in large size detectors, given that the arrival timing difference of the reflected light from different parts of the dish becomes also larger. The total time spread of the Cherenkov light using parabolic mirrors is ~1–2 ns. Consequently, the signal does not present a significant broadening, allowing us to apply a smaller integration signal window and hence, less background or noise is saved. The focal length of the reflector (distance at which the camera is placed) is also 17 m. The Point Spread Function (PSF) in each individual mirror is defined as the diameter at which 39% of the light from a point-like source is contained. This value is ~10 mm at the on-axis of the camera. In order to account for any structure deformation, at the

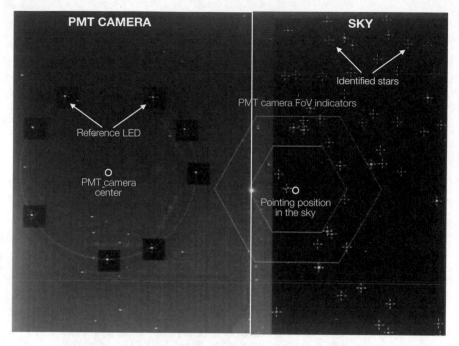

Fig. 2.12 MAGIC *Starguider camera* image

beginning of each operational night images of bright stars (at different Zd and Az) are taken with the *T-points camera*. Any deformation, and even non-optimal weather conditions as high cloudiness, can be measured quantitatively based on the PSF of the bright stars' images (see Fig. 2.13). If needed, the mirrors can be focused with the AMC system. The latter is a system composed by two actuators installed on the back of each mirror. Each actuator moves the mirrors with a precision of less than 20 μm, providing a very good pointing accuracy. The adjustment is made through Look-Up Tables (LUT) binned in Az and Zd.

2.4.1.3 Pre- and Post-upgrade Cameras

The PMT cameras of the telescopes play a key role in the overall instrument. As discussed at the beginning of this chapter, during summers of 2011 and 2012, besides upgrades on the trigger and readout systems that affected both telescopes, the MAGIC I camera suffered a major upgrade to mimic the MAGIC II camera. In this thesis, I analyzed data taken in stand–alone mode (only with MAGIC I, before 2009), in stereoscopic mode before the upgrade period and also post–upgrade data. Therefore, the former MAGIC I camera and the current one, clone of the MAGIC II camera, are described in this section (see Table 2.1).

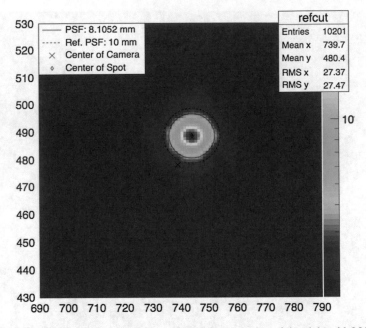

Fig. 2.13 PSF of the bright star Betelgeuse taken at the beginning of the night with MAGIC I in order to check any possible mirror misalignment. The obtained PSF (8.1 mm; red dotted circle) is inside the reference (10 mm; blue dashed circle) and therefore no adjustment is needed

Table 2.1 Comparison pre- and post-upgrade MAGIC I camera. The MAGIC II camera is identical to the post-upgrade MAGIC I and therefore, the listed features also apply to MAGIC II

	MAGIC I camera	
	Pre-upgrade	Post-upgrade
Shape	Hexagonal	Circular
Total FoV	3.6°	3.5°
Back design	Non modular	Modular
Number of PMTs	576	1039
PMT FoV	0.1° − 0.2°	0.1°
PMT QE	28%	32%
Trigger region	2.4° ⊘	2.5° ⊘

MAGIC I camera (pre-upgrade) Before 2011, the MAGIC I camera presented an hexagonal shape, covering a FoV of 3.6° (see Fig. 2.14). It was composed by 576 PMTs: 396 inner PMTs, located within a 1.2° radius from the center of the camera, and 180 PMTs in the outer part of the camera. The former, with a diameter of 1 in., covered 0.1° FoV each, while the outer PMTs covered 0.2° FoV. The trigger region corresponded to the inner zone covered only by 1 in. PMTs. These two different types of PMTs aim to achieve a good compromise between the cost and performance.

Fig. 2.14 Front (*left*) and back side (*right*) of the pre-upgrade MAGIC I camera

The Quantum Efficiency (QE) of the PMTs was around 28%. Aluminum tubes (modified version of Winton cones) were used as light concentrators to avoid losing photons in the spaces between PMTs, increasing this way the light collection efficiency. On the other hand, light collectors prevent photons coming from large angles to reach the detector, avoiding part of the NSB radiation and emission reflected by the ground. Light arriving from an incident angle larger than 40° with respect to the PMT was not collected. The total QE of the camera, accounting for PMTs, mirrors and light collectors was ∼15% in a wavelength range between 300–600 nm. Equal to the current design, the PMT used in the former MAGIC I camera had 6 dynodes and worked at a low gain to be able to observe during moonlight conditions without damaging the detector or accelerating its aging. This low-gain signal was afterwards pre-amplified. The communication between the hardware and the Counting House (CH), where signals are digitized, was performed through optical fibers. The use of optical fibers minimizes the dispersion during transmission and avoid electromagnetic pickup and most of the attenuation.

MAGIC I and MAGIC II camera (post-upgrade) The current MAGIC I camera, upgraded between 2011–2012, was a clone of the MAGIC II camera installed in 2009 (Fig. 2.15). Both present circular shape with ∼1.2 m diameter and a FoV of 3.5°. They are composed of 1039 PMTs uniformly distributed with a 0.1° FoV each one. The PMTs are grouped in 169 clusters of 7 pixels (PMTs), of which 127 clusters are completed whilst 43 are only partially equipped (outer clusters). The trigger region is 5% larger than the previous MAGIC I camera with 2.5° diameter. The PMTs are Hamamatsu R10408 with also 6-dynode system, whose QE is around 32–34% at 350 nm (Nakajima et al. 2013). The High Voltage (HV) of ∼1.25 kV is produced by a Cockroft-Walton Direct Current (DC)–DC converter. In the same way as for the former camera, the low gain of the PMTs is then compensated by a low-noise pre-amplifier. After the pre-amplification, electrical signal is also converted into optical to be able to transmit the information through 162 m long optical fibers to the CH. Post-upgraded cameras present a modular distribution that allows to extract groups of PMTs and makes the access or replacement of them easier. The pre-upgraded

Fig. 2.15 Front (*left*) and back side (*right*) of the MAGIC II camera. The current MAGIC I camera is a clone of this

camera did not have this modular design. The cameras have a plexiglass window installed in front of the light collectors to protect it from the environment conditions. There are also movable lids that prevents damage on the camera due to strong light (as Sun light) and external agents. The pre-upgraded camera was also equipped with both the plexiglass panel and lids.

After leaving the camera (through optical fibers), the signal arrives to receiver boards placed in the CH, where it is converted to electric signal again via photodiodes. From these receivers the signal is divided and sent to two subsystems simultaneously: the trigger and the readout systems.

2.4.1.4   Trigger System

The function of the trigger system consists on discriminating gamma ray-induced cascades from NSB. The total system is comprised of several steps. The following description of these steps is valid for both old and current MAGIC cameras working in stereoscopic mode. The main difference between the pre- and post- upgraded cameras is the already mentioned 5% increased trigger region in the newest device. Some discrepancies appear for mono observations, i.e. when MAGIC I operated in stand-alone mode (before 2009). Differences on that trigger system with respect to the one used nowadays are highlighted in the following when applies.

- **Level 0 (L0) trigger**: Located already in the receiver boards, this trigger releases a squared signal every time the analog signal from an individual PMT overpasses a certain amplitude threshold, the so-called Discriminator Threshold (DT). The level of this DT depends on the moonlight, being more relaxed during dark observations (non-affected by moonlight or stars in the FoV) and more conservative as higher the NSB is. A dedicated channel rate counter allows to obtain the individual pixel rate on-line (during data taking). According to this, the DT values are slightly modified automatically to keep a stable rate during variable light conditions.
- **Level 1 (L1) trigger**: The L1 trigger works over 19 overlapping hexagonal cells, the so-called macrocells, composed of 37 PMTs, one of which is blind (see

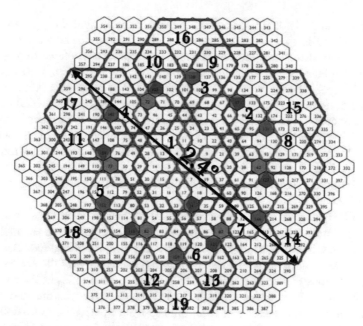

Fig. 2.16 Hexagonal L1 macrocells in the former MAGIC I camera version, each of which contains 37 PMTs. The numbers on the macrocells are the internal MAGIC identification. The trigger FoV is 2.4° diameter. Modified plot from Zanin (2011)

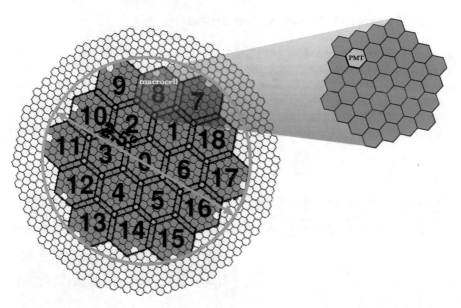

Fig. 2.17 Hexagonal L1 macrocells in the current MAGIC camera version, each of which contains 37 PMTs (one blind). The numbers on the macrocells are the internal MAGIC identification. The hexagonal shape of the PMTs is given by the Winston cones. The trigger FoV is 2.5° diameter. Modified plot from López-Coto (2015)

Figs. 2.16 and 2.17). Its inputs are the signals given by the L0 trigger from each pixel. Thus, L1 is used to find spatial and timing coincidence between closer pixels. If a number n of neighboring pixels in any macrocell, defined in MAGIC as n Next Neighbour (NN), contains a signal above the DT (i.e. the L0 issued the squared signal), the L1 trigger releases a signal. The possible n NN are n = 2, 3, 4 and 5, although during mono observation n = 4 was usual whilst currently in stereoscopic mode applies 3 NN. Therefore, to accept the trigger in a certain macrocell, a pixel that exceeds the DT must be in contact with at least other two PMTs (three in stand-alone mode observations) which also overpass the amplitude threshold. The L1 signal from each macrocell that fulfills the 3 NN criterion is processed by a Trigger Processing Unit (TPU). This is the last trigger step for mono observations.

- **Level 3 (L3) trigger**: This trigger level only applies for stereo observations. In that case, the L3 receives the output of the TPU, one signal from each telescope that triggered. If only one telescope triggered an event, it is discarded. The signals are artificially stretched to 100 ns and delayed accounting by the Az and Zd of the observations to take into account the different arrival times of the cascade in each telescope. The width of 100 ns is used to avoid loose events due to some time misalignment of the telescopes. If L1 signals from MAGIC I and MAGIC II are spaced less than 180 ns, then the event is accepted and the readout starts.

2.4.1.5 Former and Current Readout Systems

As mentioned before, one of the two signal paths goes to the readout system, responsible for the DAQ. This signal is delayed a few ns in order to wait for the trigger system response, which will determine if an event is produced by a gamma ray and hence, has to be recorded by DAQ. The readout, consisting of a Flash Analog-to-Digital Converter (FADC) in the older versions and a Analog-to-Digital Converter (ADC) in the current ones, digitizes the incoming electrical signal, which is afterwards saved

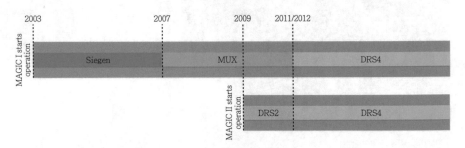

Fig. 2.18 Diagram of MAGIC readout systems and cameras along the years. The pink box represents the period in which the old camera was used, while the blue box corresponds to the new and currently used camera. The different readout system periods are labeled and depicted in different colors

as raw files. The former MAGIC I and the current MAGIC I and MAGIC II telescopes present very different readout systems, which have been updated along the years (see Fig. 2.18). In total, MAGIC has used four different readout electronics. Data from all, except the first readout system (*Siegen*), was analyzed for this thesis. I will give a description of the used readouts.

Former MAGIC readouts From February 2007, the readout system used in MAGIC was the so-called **MUX**. The PMT signals were digitized at a frequency of 2 GSample/s (2×10^9 samples per second) by FADC, which wrote that information into a ring buffer. When the trigger condition was fulfilled, a section of the buffer was saved by the DAQ. With respect to the first readout system, MUX reduced the integration time considerably, owing to the faster sampler frequency (compared to the former 300 MS/s), and so did the NSB influence. Therefore, this readout system led to an improvement in sensitivity of the ~40%. The deadtime (period in which the system is saving an event and therefore it cannot keep reading further incoming signal) with this configuration was 25 μs. MAGIC II started operation in 2009 with a readout system based on a **Domino Ring Sampler version 2** (**DRS2**) analog memory chip (while MAGIC I remained using MUX). These analog memories worked as follow: the signal coming from the receiver boards enters in a 1024 capacitor array. While the signal passes through a capacitor, the device is charged during a certain time. The switch to the next capacitor is controlled by fast switches synchronized with an external clock. Therefore, the charge in a capacitor is proportional to the time period of the clock (which is known as Domino wave). Once all the capacitors are charged, the system starts to overwrite them. The velocity sample was 2GSample/s and given the total of 1024 capacitors, the buffer longitude in DRS2 was 512 ns. For the DRS2 readout, if the trigger system accepted an event, the readout stopped the Domino wave and the charge of all the 1024 capacitors were digitized by ADC. From these 1024 capacitors, only a small fraction of them were saved by DAQ, the so-called Region of Interest (RoI). The RoI in DRS2 was 80 capacitors. Although only 80 capacitors are recorded, the system needed to read all the buffer, which led to a considerably high deadtime of 500 μs. The reader is referred to Bitossi (2009) for more information on the DRS2.

Current MAGIC readout During the major upgrade in 2011, both MAGIC I and MAGIC II readout were substituted by **Domino Ring Sampler version 4** (**DRS4**) analog memories (Sitarek et al. 2013). Besides uniforming the readout system in both telescopes, the DRS4 allows us to reduce the high deadtime. Its functionality is the same as described for DRS2. Nevertheless, instead of digitizing the 1024 capacitors after triggering, the DRS4 only digitizes and records the selected RoI. The RoI is determined attending to two points: first, if the RoI is too large, memory space is being misused with undesirable electronic noise. On the other side, if the RoI is too short, cascades with long timing development could not be entirely recorded, losing information. The RoI was reduced to 60 capacitors in the new system, which was proved to be a good agreement between both conditions. With a 33 MHz ADC, the deadtime is only 26 μs. Besides an improvement of the deadtime with respect to the DRS2, the DRS4 chips also presented low pedestal noise (0.7 photoelectron (phe) per capacitor, instead of the former 1.4 phe). Both Domino Ring Sampler (DRS) versions are temperature-dependent and therefore, observations are performed after reaching a stable electronic temperature.

Along the years and different readout configurations, the DAQ presented certain rate limitations, i.e. a maximum number of events that can process per second. The main limitation is due to the writing speed on the disk. If the rate is too high, instabilities, unnecessary loss of time due to the deadtime and, more important, crashes on the system can happen. The recovery from a DAQ crash takes several minutes, during which data taking is not possible. An increase of rates can arise if the DT of the L0 trigger is too low, allowing NSB events to fulfill the amplitude threshold condition and increasing this way the probability of accomplishing the NN criterion with just background events. This very low DT scenario and consequences also apply to After Pulses (AP). Even if the DT is set according to the observational conditions, a very bright star in the FoV during the automatic repositioning of the telescopes due to a GRB alert (see Sect. 2.4.1.6) or artificial lights pointing to the camera (like car flashes) can disturb the DAQ system. To avoid this, besides the automatic tunning of individual pixels DT under bright observational conditions (Sect. 2.4.1.4), a *rate limiter* is installed to stop momentarily the trigger system if the stereo rate exceeds 1 kHz. During moon observations, the maximum rate that the DAQ could handle was 600 Hz.

All the recorded events by DAQ have a time stamp proffered by a Rubidium clock. The Rubidium clock, with a precision of 3×10^{-11} per second, provides the absolute time. It is in turn synchronized with a Global Positioning System (GPS) with a precision of ns.

2.4.1.6 Other Subsystems

On-line systems These are those systems that work during data taking. Here there are encompassed the GRB monitoring alert system and MAGIC OnLine Analysis (MOLA).

- **GRB monitoring alert system**: It monitors the Gamma-ray Coordinate Network (GCN)[1] in order to alert from a possible GRB event. The program evaluates the GCN alert in terms of observability (zenith range, distance to the Moon and Sun, uncertainties on the given GRB coordinates). If the GRB alert accomplishes the observational conditions, the CC takes control over the telescopes and move them automatically to the GRB position in order to lose the minimum possible observation time. The repositioning of the telescopes for the maximum Az possible difference of 180° takes only 20 s, thanks to their light structure.
- **MOLA**: It is a multithreaded C++ program that gives an on-the-flight estimation of the gamma-ray emission when observing a source (Tescaro et al. 2013). It provides θ^2 plots, light curves and skymaps (see Sect. 2.4.3) of the target. This tool is specially interesting when observing a transient system, where flares can occur.

---

[1] https://gcn.gsfc.nasa.gov/.

Monitoring weather systems: The main goal of these subsystems is to control the weather conditions in order to evaluate if observations can be performed or to correct data afterwards during analysis if needed. They are: weather station, pyrometer, Light Detection And Ranging (LIDAR) and AllSky camera.

- **Weather station**: The weather conditions such as humidity, wind speed and direction, temperature and pressure are given by a Reinhardt 5MW weather station, located in the roof of the CH. MAGIC telescopes can operate only if safety conditions are fulfilled.
- **Pyrometer**: This instrument measures the temperature of the sky and provides an atmospheric transparency estimation. It is only installed in the dish of MAGIC I and points to the same direction than the telescope. It measures IR radiation (in the 8–14 μm band) that fits to a blackbody spectrum, obtaining this way the temperature. The measured temperature increases if the sky is cloudy, because it reflects radiation from the ground. Thus, an estimation of the cloudiness (higher cloudiness implies lower transparency) is given by:

$$\frac{T_{low} - T_m}{T_{low} - T_{up}} \tag{2.7}$$

  T_{low} and T_{up} correspond to the temperature of the sky at its worst and best conditions, respectively, which are set to $T_{low} = 250$ K and $T_{up} = 200$ K. T_m is the measured temperature by the pyrometer.
- **LIDAR**: The LIDAR, located in a dome on the CH roof, is equipped with a 5 mW Q-switched pulsed laser, a 60 cm diameter aluminum reflector with 1.5 m focal length, a Hybrid Photo Detector (HPD), a robotic equatorial mount and a computer that uses FADC cards to digitize the signal from the HPD. The LIDAR shoots the laser at a position shifted by 3° from the observing source, taking care not to disturb MAGIC data taking, to measure the transparency of the atmosphere. It fires 50000 laser shots with a frequency of 300 Hz and hence, each data run takes around 3 min. A new run is started every 5 min. The pulsed light from the laser is backscattered by the clouds and aerosols on the sky. The transparency is measured as a function of the arrival time distribution of the backscattered photons. It can provide transmission estimation at different altitudes of 3, 6, 9 and 12 km (Fruck et al. 2014).
- **AllSky camera**: Monochrome AllSky-340 SBIG camera placed in the roof of the CH. It points to the Zd and provides an image of the sky every 2 min. It has complete FoV of 360° in Az and almost 90° in Zd (see Fig. 2.19). These images are monitored online during observations.

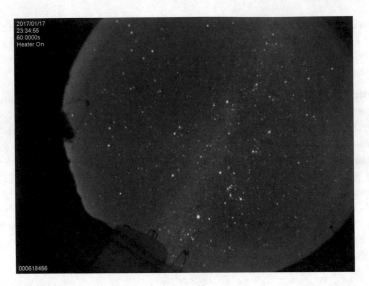

Fig. 2.19 MAGIC AllSky-camera image

2.4.2   Data Taking

The low gain PMTs used in the MAGIC cameras allow us to observe under dark (without the Moon presence) and moderate moonlight conditions. The maximum dark duty cycle in a year corresponds to an 18% ($\sim$1500 hr/year), of which $\sim$65% is observed and the rest is lost because of technical problems or bad weather conditions. Normally, observation during full Moon nights and around are stopped. Nevertheless, by observing also during decent or strong Moon presence, this duty cycle increases considerably by $\sim$40%. To do so, small hardware modifications are applied to the standard setup, like decreasing the HV or using UV-filters, which deal with the higher NSB. The mean pixel DC is proportional to the NSB level: the higher the NSB, the higher the DC is. Thus, the NSB of certain observations can be estimated by comparing the corresponding median pixel DC and the median pixel DC under dark optimal-weather conditions (which provides the reference level, NSB_{dark}). Given that the distribution of photons reaching the PMT photocathode follows a poissonian distribution (see Appendix B), an increase of the NSB light will produce a raise of the image noise proportional to $\sqrt{NSB}$. A raise of the incoming light will also give rise to higher AP rate. This situation leads to an increase of the accidental rate in the L0 trigger, i.e. more NSB-induced cascades fulfill the trigger threshold. This has a major effect on the low energy cascades that can be embedded on the high noise. To suppress these undesirable events, stronger cuts are applied during the data analysis (Sect. 2.4.3). Attending to safety reasons, with the standard HV of $\sim$1.25 kV, the maximum sky brightness under which MAGIC can observed is $12 \times \text{NSB}_{dark}$. By reducing this HV value a factor of $\sim$1.7 (the so-called reduced HV), the observations

Fig. 2.20 MAGIC
UV-filters used during full
Moon observations

can be extended up to $20 \times \mathrm{NSB}_{dark}$. This reduced HV performance allows MAGIC to observe up to 90% of the Moon phase. Observations can be carried out during full Moon if UV-filters, as those shown in Fig. 2.20, are installed. These filters preserve large fraction of the Cherenkov radiation, allowing transmission in the UV band (with a peak at 330 nm), while blocking longer wavelength. The MAGIC performance under moonlight is reported in MAGIC Collaboration et al. (2017).

2.4.2.1 Observational Pointing Modes

There are two possible pointing modes with MAGIC, the *ON/OFF mode* and the so-called *wobble mode*. Given that both pointing modes were used during observations of sources included in this thesis, I will explain both of them. Nevertheless, the current standard pointing is the *wobble mode*.

- **ON/OFF mode**: The source is tracked at the center of the camera. The *ON region* (where the signal from the source is expected) and the *OFF region* (background signal, used in the significance computation, see Sect. 2.4.3.9) are observed separately. Here, the background sample is recorded under same conditions (same epoch, zenith angle and atmospheric conditions) as for the *ON data* but with no candidate source in the FoV.
- **Wobble mode**: In this case, MAGIC points at two or four different positions situated $0.4°$ away from the source to evaluate the background simultaneously (Fomin et al. 1994). Each of the positions are usually observed in time slots of 20 or 15 min (the

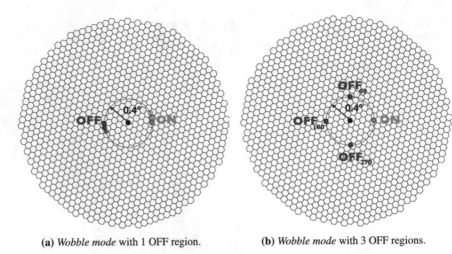

(a) *Wobble mode* with 1 OFF region.   (b) *Wobble mode* with 3 OFF regions.

Fig. 2.21 Scheme of the *wobble* pointing mode. The black circle corresponds to the center of the camera, located at 0.4° from the source (green circle). As the source moves in the sky, it remains all the time placed at this distance from the center of the camera, giving rise to a circular movement around the camera center. The background can be simultaneously taken from one region (*left plot*) or three regions (*right plot*). In the former, the OFF region (red point) is all the time situated at 0.4° from the center of the camera, at an opposite direction from the source. If 3 OFF regions are selected, the background is evaluated in regions separated the same distance one from the other, all of them at 0.4° from the camera center. The subindex of the OFF regions determines the angle at which they lay in the imaginary circle formed by the ON source movement. These angles (90° 180° 270°) are given counting counterclockwise from the ON region. Plots taken from López-Coto (2015)

so-called **runs**). With this mode, exactly same conditions of the OFF (background) data are achieved and, in turn, observational time is saved since no dedicated OFF observations are needed. To calculate the significance, one or three OFF regions can be selected (see Fig. 2.21). In the former case, the OFF region will be taken at the opposite position of the camera with respect to the source (see Fig. 2.21a). However, three OFF sources are usually encouraged to provide a better background estimation and therefore, most reliable significance. With this pointing mode, there is a reduction of systematic effect produced by different weather or NSB level conditions with respect to the *ON/OFF mode*. Nevertheless, there are also some disadvantages. The main one is a decrease on the gamma-ray detection efficiency due to the shift of the source. Because the telescopes are pointing at 0.4° away from the target, some fraction of the EM cascades lay outside of the trigger region (in mono, this fraction reaches 15–20%). On the other hand, the camera presents inhomogeneities due to the different gain and electrical noise of the PMTs or dead pixels along the trigger region. Therefore, systematic errors in the background estimation arise from the fact that ON and OFF regions are not taken from the same part of the camera, leading to an overestimation or underestimation of the signal.

2.4.3 Data Analysis

Data analysis is performed with the standard MAGIC analysis software called MAGIC Analysis and Reconstruction Softward (MARS) (Zanin et al. 2013). It makes use of a collection of object-oriented C++ programs and ROOT[2] libraries and classes. MARS converts the raw FADC/ADC counts stored by DAQ into processed high-level data. The goal of the analysis is to determine whether an event was generated by a gamma ray or a hadron (γ/hadron separation) as well as to obtain the energy and direction in case of gamma ray-induced showers. In this section, I will give a review of all steps performed by MARS, depicted in Fig. 2.22 and listed below:

- **Calibration of the signal** into phe, performed by CALibrate Light Signals and Time Off-sets (CALLISTO) (for FADC and DRS2 readout data) or Simple, Outright Raw Calibration; Easy, Reliable Extraction Routines (SORCERER) (for both DRS2 and DRS4 data).
- **Image cleaning and Hillas parameters calculation** computed by the program `Star`.
- **Stereo image parameters** with the software `Superstar`.
- **Train of the Random Forest (RF)** for the γ/hadron separation, **produce the LUTs** for the energy reconstruction and **compute *disp* parameters** for the arrival direction. This training, carried out by `Osteria` (mono) or `Coach` (stereo), needs simulated Monte Carlo (MC) gamma-ray events and a data sample of real background data (observations with no gamma ray-emitter in the FoV).
- **Apply RF and LUTs** to the real data to obtained *hadronness*, reconstructed energy and arrival direction. Also applied to the MC data. The program used for this task is `Melibea`.
- **Computation of signal significance** with `Odie`, **skymaps** with `Caspar` and **spectra and ligth curves** with `Fluxlc` or `Flute`.

2.4.3.1 Monte Carlo Simulations

In order to reconstruct the energy of the primary gamma ray and its arrival direction, IACTs need to use MC simulations. In MAGIC, MC gamma rays are simulated in two ways: as *ringwobble* MC and as *diffuse* MC. The former simulates a ring of 0.4° radius (with a width of 0.1°) from the camera center, accounting for the 0.4° offset used in the standard *wobble mode*. It is used for the analysis of point-like sources. The latter is applied for the study of extended sources or sources shifted from the nominal position. In this case, diffuse gamma rays are simulated covering a circle of 1.5° radius. Figure 2.23 shows a schematic views of both types of MC. In both cases, separated MC data are available for low Zd range (5°-35°), medium Zd range (35°-50°), high Zd range (50°-62°) and very-high Zd range (62°-70°). For more

---

[2]https://root.cern.ch/.

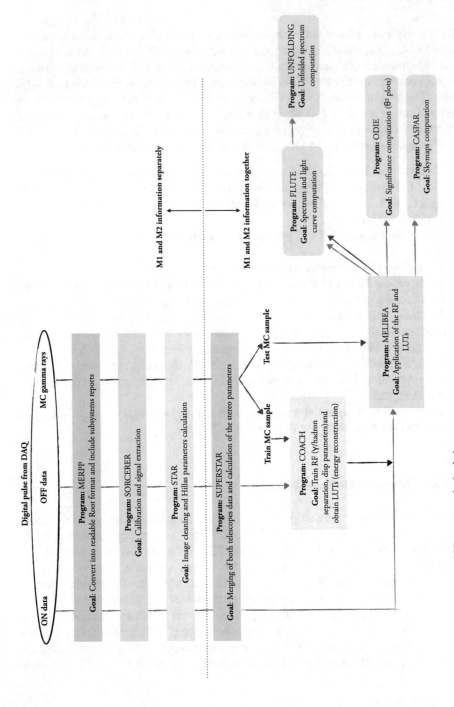

Fig. 2.22 Flowchart of the MAGIC stereo analysis chain

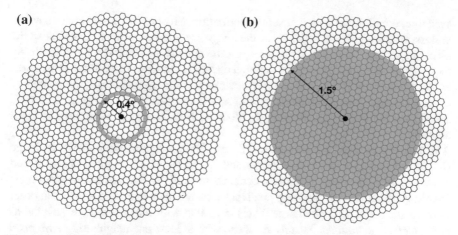

Fig. 2.23 Schematic view of the *ringwobble* (*left*) and *diffuse* (*right*) MC. The green area corresponds to the region in which gamma rays are simulated. For the *ringwobble* MC a width of 0.1° is used. López-Coto (2015)

information of MC simulations with MAGIC, the reader is referred to Majumdar et al. (2005).

2.4.3.2 Signal Pre-processing

To be able to analyze the raw data coming from the FADC or ADC (counts as a function of time) with the standard MARS software, one needs to convert it into ROOT format. This conversion is made by MERging and Prepoccesing Program (MERPP), which also attaches other subsystems reports to the ROOT files.

Once ROOT files are available, the calibration of the counts into phe is carried out by SORCERER. In the past, another program called CALLISTO was used. As mentioned in Sect. 2.4.1.5, once the trigger is accepted the DAQ stored the information of a certain RoI per pixel, which was 80 capacitors before 2009 and currently 60 capacitors. Each capacitor contain the charge of the ADC in a time of 1/(sampling speed) ns and hence, a total of 30 ns are recorded now from the DRS4. The signal pre-processing goal is to achieve the charge (total signal) and arrival time of each event from the pixels. After the signal extraction, the measured charge is calibrated to obtain phe through the F-Factor method (Mirzoyan 1997) as explained in Appendix B.

2.4.3.3 Image Cleaning and Hillas Parameters Calculation

After the calibration, the images are cleaned and parametrized by the Star program. Although after the signal pre-processing, charge and arrival time for each PMT is available, not all pixel contain useful information. Most of them only contain noise, useless for the signal analysis. Thus, the *image cleaning algorithm* aims to keep the pixels in which Cherenkov photons from the shower produced signal, discarding those pixels that, below a certain calibrated signal timing and amplitude thresholds, do not contain useful information of the shower image. In this thesis, data with two different image cleaning methods were used: the absolute image cleaning (Aliu et al. 2009b) and the sum image cleaning (Lombardi 2011). The latter is the currently used algorithm.

Absolute image cleaning This method discriminates pixels according to their charge (in phe), dividing between *core pixels* and *boundary pixels*. A pixel is considered as *core pixel* if its charge is above a certain threshold Q_c. Moreover, this pixel must have at least another neighboring pixel that accomplishes this threshold too, in order to avoid that random high charge pixels unrelated with the cascade are selected as *core pixels*. The *boundary pixels* are those with at least one neighboring *core pixel* and that is above a Q_b threshold. The selection of the charge thresholds, Q_c and Q_b, is important for further analysis. In order to detect the lowest energy gamma-ray cascades, these threshold should not be restrictive. However, very relaxing thresholds can in turn allow noise-induced images to overpass this condition. To solve this, the absolute image cleaning makes use of the signal arrival time, letting the signal charge thresholds be lower with less risk of accepting non gamma-ray events. The information used is based on the fact that Cherenkov radiation lasts a few ns, much less than the arrival time for NSB. The applied conditions are the following: the *core pixel* cannot differ from the mean arrival time obtained from all *core pixels* more than Δt_c, whilst the *boundary pixels* time cannot differ more than Δt_b from the *core pixel*'s arrival time. The timing thresholds are, for all periods and performances, $\Delta t_c = 4.5$ ns and $\Delta t_b = 1.5$ ns.

The threshold values change according to the observational conditions, like the level of moonlight. It also depends on the camera performance and hence, different values were applied for pre- and post-upgraded camera. In this thesis, I analyzed data under dark conditions for pre-upgrade period and under dark and all moonlight levels for the post-upgrade period. The corresponding values used in the image cleaning of the data are listed in Tables 2.2 and 2.3. As mentioned in Sect. 2.4.3, for the complete analysis of the data, MC data and a real background sample data are needed, both of which need to mimic same observational conditions are those for the source data. This implies that for Moon analysis, one needs appropriate MC and background data at the same moonlight levels than the observations. MC simulations are only computed for dark conditions and therefore, artificial noise is injected in the files before the image cleaning performed at the `Star` level. In the case of background sample, one can introduce artificial noise as well or use data taken under same moonlight conditions. The level noise that has to be inserted is given by the mean pedestal value (and its Root Mean Square (RMS)) of the so-called interleaved pedestals. These runs are randomly triggered events recorded during observations in order to evaluate the NSB. Thus, for each moonlight level (divided according to the NSB or DC -see Table 2.3), we select the interleaved pedestal run that shows the most similar mean DC conditions than the mean DC of our data observations. Table 2.3 reports the currently standard

Table 2.2 Image cleaning levels for pre-upgraded cameras. Standard HV value corresponds to ~1.25 kV

	NSB (× NSB$_{dark}$)	Equivalent DC [μA]	HV	Q_c [phe]	Q_b [phe]	Δt_c [ns]	Δt_b [ns]
MAGIC I	1–2	<2	Standard	6	3	4.5	1.5
MAGIC II	1–2	<2	Standard	9	4.5	4.5	1.5

Table 2.3 Image cleaning levels for post-upgraded cameras. Standard HV value corresponds to about 1.25 kV, while the reduced one is a factor ~1.7 lower

NSB (× NSB$_{dark}$)	Equivalent DC [μA]	HV	Q_c [phe]	Q_b [phe]	Pedestal mean factor [phe]	RMS factor [phe]
1–2	1.1–2.2	Standard	6	3.5	–	–
2–3	2.2–3.3	Standard	7	4.5	3.0	1.3
3–5	3.3–5.5	Standard	8	5	3.5	1.4
5–8	5.5–8.8	Standard	9	5.5	4.1	1.7
5–8	3.2–5.2	Reduced	11	7	4.8	2.0
8–12	5.2–7.8	Reduced	13	8	5.8	2.3
12–18	7.8–11.6	Reduced	14	9	6.6	2.6
8–15	2.2–5.0	Standard with UV-filters	8.0	5.0	3.7	1.6
15–30	5.0–8.3	Standard with UV-filters	9.0	5.5	4.3	1.8

noise levels according to the NSB and the corresponding image cleaning. The latter becomes more restrictive as higher the NSB is.

Sum image cleaning This algorithm is also based in the previously given definition of *core* and *boundary pixels*. However, it uses the compactness of the image to obtain more information. With this additional information the charge thresholds can be decreased, which benefits the low energy cascades. The sum image cleaning looks for N (N = 2, 3 or 4) neighboring pixels whose summed charge is above a threshold Q_c within a time window given by t_c. Before summing, the signals are "clipped" if they are above a certain value to avoid APs. The value of Q_c and t_c depends on the selected N, as presented in Table 2.4. The *boundary pixels* are those adjacent with at least one *core pixel*, with an amplitude above a fixed value of 3.5 phe and a arrival time difference with respect to their neighboring pixel of <1.5 ns.

Hillas parameters After the image cleaning, an ellipse is fit to the surviving pixels and the momenta of this fit (up to second order) are the so-called Hillas parameters used in the MAGIC analysis (shown in Fig. 2.24). The parametrization information of

Table 2.4 Sum cleaning
image parameters

Topology	Q_c [phe]	t_c [ns]
2NN	10.8	0.5
3NN	7.8	0.7
4NN	6	1.1

Fig. 2.24 Schematic view of
the Hillas parameters

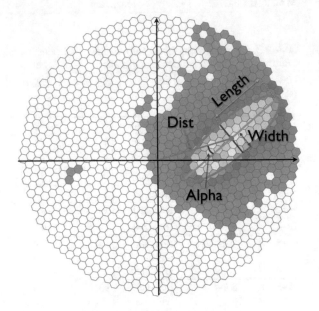

each event is stored in ROOT files for both telescopes separately. The main parameters
calculated are:

- **Size**: It corresponds to the sum of the charges in phe of each surviving pixel. The
  size is correlated to the energy of the primary gamma ray if the event is contained
  in the Cherenkov light pool of radius ~120 m.
- **Length**: Longitude of the major semi-axis of the ellipse. It is related with the
  longitudinal development of the cascade.
- **Width**: Longitude of the minor semi-axis of the ellipse. It is a measurement of the
  lateral development of the cascade.
- **Conc(N)**: Fraction of the image charge contained in the N brightest pixels. It gives
  the compactness of the image, which for EM cascades is larger than for hadronic
  showers. The used value is Conc(2).

  Some of the Hillas parameters are **source-dependent**. This means that although
they represent the physical features of the showers, these depend on the source
position.

- **Dist**: Angular distance between the position of the source and the center of gravity
  of the image. The larger the dist value, the larger the impact parameter of the
  shower in the ground.

- **Alpha**: Angle between the major axis of the ellipse and the imaginary line connecting the source position and the center of gravity of the image. During ON observations, the source position corresponds to the camera center (for the *wobble mode* this is shifted $0.4°$) and therefore EM cascade should point to the camera center. Thus, gamma ray-induced showers have smaller alpha angle than the hadronic ones.

Time-dependent parameters are also useful to discriminate between EM and hadronic showers, given that the former develop faster (~ 3 ns compared to the ~ 10 ns):

- **Time RMS**: RMS of the arrival time of the surviving pixel, which is smaller for gamma ray-induce cascades.
- **Time gradient**: Slope of the linear fit applied to the arrival time projection along the major axis, which gives the direction of the shower development. EM showers are expected to have positive development (from close to the camera center to outside).

Other parameters are used to estimate the **image quality**. Thus, very noise images or images not well-contained in the camera can be discarded.

- **LeakageN**: Fraction of the shower light contained in the N outermost rings of PMTs of the camera (usually, $N = 1$). This parameter measures how much image is contained in the camera.
- **Number of Islands**: Number of surviving pixels after image cleaning non-related with the image event.

Finally, there are the so-called **directional parameters**. They are used to differentiate between the head (top of the cascade) and tail (bottom of the cascade). Atmospheric showers present higher charge in the head part, since particles in the top have higher energies.

- **Asymmetry**: Direction of the line between the center of gravity of the image and the pixel with the highest charge. The EM cascades present positive asymmetry, i.e. pixels with the highest charge are located close to the source position.
- **M3Long**: Following the same criterion as the asymmetry parameter, M3Long is the third moment of the image along its major axis.

2.4.3.4 Data Quality Cuts

After running the program `Star`, the ROOT data files contain all the information (image parameters) of each event to proceed on the calculation of the *significance* or flux. However, before continuing, data quality cuts are necessary to guarantee reliable results. Data quality of the events is disturbed by technical problems or bad weather conditions and can be estimated by using different indicators. One of them is the **Rate of events** (in Hz) in each subrun (data sample of ~ 2 min) for a low

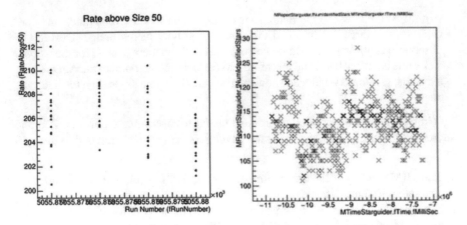

Fig. 2.25 Example of two plots used in the data quality selection. *Left panel*: Rate plot as a function of the subrun, taken from Cygnus X-3 observations with MAGIC II applying a *size cut* of 50 phe. *Right panel*: Number of identified stars by the Starguider camera for the same night. The high number of detected stars reveals very good quality weather conditions

size cut of 50 phe (Fig. 2.25). Subruns with unreliable high rate, due to accidental events, can be discarded. One must take into account that the rate is Zd dependent, as the absorption is higher at larger Zd angles. This can be corrected (up to $\sim 50°$) by multiplying the rate by $\cos^{-1}(Zd)$. During a good weather condition night, the rate should stay stable and hence, we normally accept events with rates that differ $\pm 15\%$ from the mean rate. Because of the higher reflectivity on the MAGIC II mirrors, the rate for this telescope tends to be slightly higher than in MAGIC I. If the observations are taken under very cloudy conditions, the rate will be unexpectedly low and so, data are rejected as non-optimal for analysis. A proper estimation of the sky coverage can be made with the *cloudiness parameter*, given by the pyrometer installed in MAGIC I (see Sect. 2.4.1.6). For this thesis, and as usually done, I considered *bad data* those with a cloudiness above 40%. The information stored from the Starguider camera is also useful to determine if the data should be classified as *bad*. Given that the Starguider compares the observed FoV with a star catalog, if only a small fraction of stars are recognized that would imply high cloudiness or humidity (see Fig. 2.25).

2.4.3.5   Stereo Image Parameters Calculation

After the data quality selection, the `Superstar` program merges the information from both telescopes. The new stereo image parameters obtained are the necessary tools for the energy and direction reconstruction performed in further steps. The most important parameters are listed below. Schemes of some of them are presented in Figs. 2.26 and 2.28.

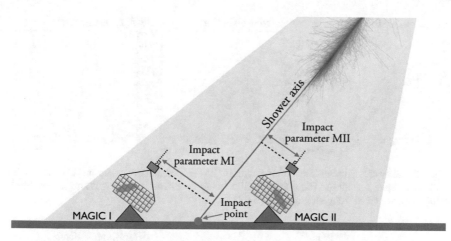

Fig. 2.26 Geometry of stereo event, where the impact parameters for both telescopes, impact point and shower axis are highlighted

- **Shower axis**: This axis can be calculated as the crossing of the enlarged major axes of the two images of the telescopes when they are superimposed in the same camera plane (Fig. 2.28b). This is the so-called *crossing point* (Aharonian et al. 1997; Hofmann et al. 1999). This method cannot be used in mono observations, for which the *Disp method* is applied. The latter, more robust than the *crossing point*, is as well used for the current MAGIC stereo observations (see Sect. 2.4.3.7).
- **Impact point**: The point in the ground that the shower axis reaches. It is determined by the crossing of the enlarged major axes of the image shower in each of the telescopes, taking into account their position (see Fig. 2.28a).
- **Impact parameter**: Perpendicular distance in the camera between the pointing direction and the shower axis. There is one impact parameter per telescope. See Fig. 2.26.
- **Shower maximum height** (H_{max}): The altitude at which the number of particles in the cascade is maximum (H_{max}) is determined, once the shower axis is known, with the angle at which the image of the center of gravity is viewed in each telescope. As shown in Sect. 2.2, H_{max} depends on the energy of the primary particle: the higher this energy is, the closer to the ground the cascade develops and hence, H_{max} is smaller. For the γ/hadron separation, parameters such as width or length are usually more relevant. However, at lower energies, H_{max} plays an important role helping to differentiate the nature of the shower (See Fig. 2.27).
- **Cherenkov radius** (R_C): Radius of the Cherenkov light pool produced by an electron with the bremsstrahlung critical energy of 86 MeV at H_{max}. This parameter is obtained from MC simulations.

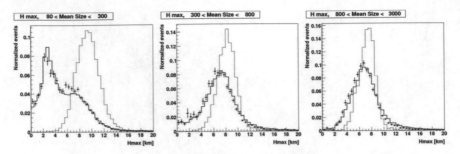

Fig. 2.27   Shower maximum distribution height for EM (dotted lines) and hadronic cascades (solid line) for different *size* cuts (in phe). The MC distribution for hadronic showers is also shown (black points). Images taken from Aleksić et al. (2012b)

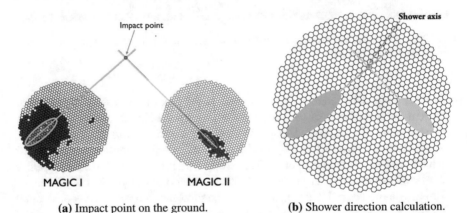

(a) Impact point on the ground.   **(b)** Shower direction calculation.

Fig. 2.28   Stereo parameters calculation

• **Cherenkov photon density**: Density of Cherenkov radiation produced by an electron with the bremsstrahnlung critical energy of 86 MeV at H_{max}. This parameter is obtained from MC simulations.

2.4.3.6   γ/Hadron Separation

As mentioned before, even for strong sources as the Crab Nebula, the ratio between hadronic and EM showers is 1000:1 at hundred GeV. This is the reason why an optimal γ/hadron separation is key in the analysis. To perform this discrimination, we use a RF, a multi-dimensional classification algorithm based on decision trees (Albert et al. 2008). In order to train the RF on how the gamma-ray events look like compared to the hadronic ones, the algorithm uses two inputs: MC simulated gamma rays and real background data (with no gamma-ray emitter, to avoid misleading the training of the RF algorithm). Both of them need to mimic the observational conditions under which the source data were taken, attending basically to weather conditions, moonlight and zenith range. The MC set applied here has to be different for the one used later on the flux calculation, in order to not bias the analysis. Thus, the entire MC sample is divided into two sub-sets, a *train sample*, used for the γ/hadron separation, and a *test sample*, used to calculate the collection area and migration matrix computation.

The RF tree starts with the whole sample of events (containing both gamma rays from MC and hadrons from background data), which provides a reliable image of the real scenario when observing. A pre-selected set of P parameters (such as *size*, *length*, *arrival time*, etc.) are used to discriminate between gamma rays and hadrons. The γ/hadron separation is obtained by dividing the initial sample into two subsamples of events, gamma rays and hadrons, based on optimized cuts of one randomly selected P parameter at a time. The optimization of the cuts is based on the minimization of the *Gini coeficient* (Gini 1921):

$$Q_{Gini} = 4 \frac{N_\gamma}{N} \frac{N_h}{N} \tag{2.8}$$

where N is the total number of events, N_γ is the number of gamma rays and N_h is the number of hadrons. The classification selects another parameter randomly and the subsequent division into gammas and hadrons takes place. If one of the subsamples contains only gamma rays or hadrons, the separation process stops in that branch. To evidence the discrimination, if the events from this subsample belongs to the gamma-ray population, they are assigned a 0, whilst if they are hadrons the assignation is 1. The training of the RF grows up to a limit of n trees, which in MAGIC is usually $n = 100$. In Fig. 2.29 there is a graphical view of the RF classification.

This trained RF is afterwards applied to real data with the `Melibea` software. Each event of the data has to pass through all the trees previously trained, which allows to classify it into gamma ray or hadron. To quantify how likely an event is a gamma ray or hadron, each event is assigned a *hadronness* value ranging from 0 to 1 (closer to 1 implies hadron-like event). The final *hadronness* value, h, of each event is determined by the mean of the *hadronness* assigned to all the trees during the training, h_i:

$$h = \sum_{i=1}^{n} \frac{h_i}{n} \tag{2.9}$$

2.4.3.7 Arrival Direction Reconstruction

The *crossing pointing method* (explained in Sect. 2.4.3.5) can be used in the reconstruction of the arrival direction of the primary gamma ray. However, for stand-alone observations a more sophisticated method, based on MC simulations, is necessary. This method, proposed by Fomin et al. (1994) and revised by Lessard et al. (2001), is the so-called *Disp method*, which is currently used in stereo observations as well.

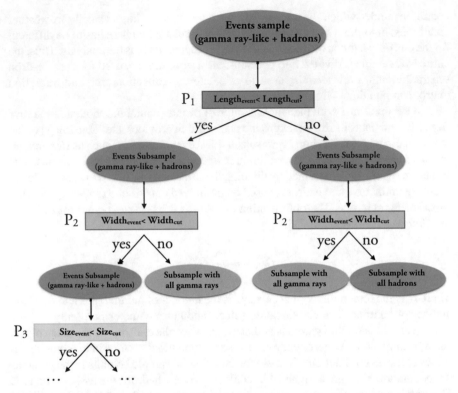

Fig. 2.29 Scheme of the RF classification. The P_i parameters are selected randomly. Once all events in a subsample belong to the gamma-ray or hadron population, the division stops. The cuts applied are optimized with the *Gini coefficient*

The *Disp method* works as follows: given that the elliptical image obtained in the camera is a projection of the real EM shower, the major axis of it represents the incoming direction of the cascade in the camera plane. Therefore, the source position has to be on this axis separated a certain distance, known as *disp*, from the center of gravity of the image. Currently, the *disp parameter* is calculated (both in mono and stereo analysis) using a method based on a RF algorithm. The RF is trained in a similar way as it is done for the γ/hadron separation. Given that the *disp parameter* is known for simulated gamma rays, the MC events pass through n number of decision trees to get a correlation between the *disp* and a set of parameters. In this case, the optimal cuts are those which minimize the variance of the *disp parameter* in each division, instead of the *Gini coefficient*. Among the previously presented image parameters (Sect. 2.4.3.3), source independent variables are important in the RF training to avoid bias, as for example the time gradient.

In the case of stereoscopic observations, given that we have one image per camera, there are four *disp distances* or estimations on the source position (Fig. 2.30b). To calculate the accurate arrival direction, we compute the distances between these four

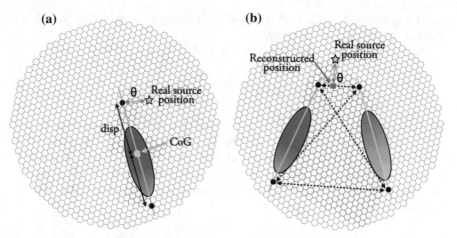

Fig. 2.30 *Disp method* for stand-alone (*left*) and stereo observation (*right*)

possible positions and select the smallest one. The reconstructed arrival direction is the average of the two closest positions weighted with the number of pixels of each image. If none of the distances are smaller than 0.22° the algorithm does not accept the reconstruction as valid.

The *Disp method* brings a new parameter in the analysis, the angular distance θ (see Figs. 2.30a, b). It corresponds to the angular distance between the true position of the source and the reconstructed one. Based on this parameter, we obtain the significance plots, the so-called θ^2 plots, which will be discussed in Sect. 2.4.3.9. The introduction of the RF in the determination of the *disp distance* led to an important improvement in the angular resolution, which is translated into a 20–30% better sensitivity (Aleksić et al. 2012b).

2.4.3.8 Energy Reconstruction

To estimate the energy of a primary gamma ray, two methods were applied, accounting for observations performed in stand-alone or stereo mode. For mono observations, the reconstructed energy is obtained in a similar way as the *hadronness* or arrival direction, by means of a RF. Since with MC data the true energy of the event (E_{true}) is known, the RF is trained with a pre-defined set of parameters selected randomly in each step, whose optimal cut is that which minimizes the variance of E_{true}.

To reconstruct the incident energy in stereoscopic observations, we make use as well of MC data but through LUTs. In a 2-dimensional histogram binned in *size* and *impact parameter/R_C* (where R_C is the Cherenkov radius), each slot is filled with the E_{true} and its corresponding RMS of each event for MAGIC I and MAGIC II, separately. The estimated energy (E_{est}) of an event is obtained with the average of the corresponding E_{true} for each image of both telescope weighted by the RMS of each bin and corrected by cos(Zd).

The **energy bias**, relative error between the E_{est} and E_{true}, is defined as:

$$E_{bias} = \frac{E_{est} - E_{true}}{E_{true}} \tag{2.10}$$

The E_{bias} for an energy bin is obtained by fitting a Gaussian to all E_{bias} for each individual events included in the bin. The E_{bias} of the system is the average of all those Gaussians. The E_{bias} keeps around zero for energies >150 GeV at low Zd range and >200 GeV at medium range. The relative error between the estimated and true energy increases considerably at lower energies due to the energy threshold (see Sect. 2.4.3.10).

2.4.3.9 Signal Extraction

To compute the significance of the signal from the source, it is used the distribution of events as a function of the squared angular distance from the real source position and the reconstructed one, θ^2. The program responsible of calculating the θ^2 distance and compute the significance is `Odie`. Assuming that the camera is homogeneous close to the center where we point to the source, the background events (N_{off}) should follow an homogeneous distribution all over the θ^2 range, whilst we expect gamma-ray events (N_{on}) to peak at small θ^2 values, i.e. nearby the source. The events used in this distribution are those which survived the analysis cuts, such as *size*, *hadronness* or a specified Zd. Although these cuts can be optimized on independent and known sources, we usually apply standard cuts for the MAGIC analysis. The latter are divided into three categories attending to the energy range in which the bulk of events are expected to lay: Low Energy (LE) with $E_{thr} = 100$ GeV (where E_{thr} is the energy threshold, see Sect. 2.4.3.10), Full Range Energy (FR) that provides a medium-to-high energy range with $E_{thr} = 250$ GeV, and HE with $E_{thr} = 1$ TeV. The optimized cuts for each range varied from pre- to post-upgraded period. Both set of cuts can be found in Table 2.5.

To determine the **significance** of the signal, we make use of the two distributions N_{on} and N_{off} separately (see Fig. 2.31a). The former, N_{on}, is obtained from the real data. These events are not purely gamma rays, but include surviving events from electrons/positrons in hadronic cascades, gamma-like hadrons or diffuse gamma rays (whose contribution is larger in galactic sources). The background contribution, given by N_{off}, is obtained differently according to the pointing mode: in the case of *ON/OFF mode*, these events are get from the dedicated OFF observations in a FoV with no gamma-ray candidate and with same observational conditions. For *wobble mode*, the N_{off} is obtained directly from the same source observations, at the same distance from the center of the camera than the source but at different direction (more information in Sect. 2.4.2.1). The excess events, N_{exc}, are determined by the difference between N_{on} and the scaled N_{off}:

Table 2.5 Standard MAGIC analysis cuts attending to the energy range for periods before (pre-) and after (post-) the upgrade of 2011–2012

	Energy range	E_{thr} [GeV]	θ^2 [deg^2]	Hadronness	Size in M1 [phe]	Size in M2 [phe]
Pre-upgrade	Low energy	100	<0.026	<0.28	>55	>55
	Medium-to-high energy	250	< 0.01	<0.16	>125	>125
	High energy	1000	<0.01	< 0.17	>300	>300
Post-upgrade	Low energy	100	<0.02	<0.28	>60	>60
	Medium-to-high energy	250	<0.009	<0.16	>300	>300
	High energy	1000	<0.007	<0.1	>400	>400

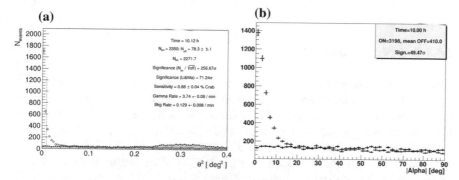

Fig. 2.31 *Left panel*: θ^2 distributions for the Crab Nebula data with medium-to-high energy cuts (see Table 2.5). The red points correspond to ON events and the black points to OFF events, taken in *wobble mode*. *Right panel*: Alpha distribution for a Crab Nebula sample taken with the stand-alone mode with MAGIC I

$$N_{exc} = N_{on} - \alpha \cdot N_{off} \qquad (2.11)$$

where the scale factor α is the normalization between N_{on} and N_{off}, given that they are usually not observed the same amount of time. The value of α is obtained by the fraction of ON events over the OFF events in a region far away from the expected signal region in the θ^2 plot, normally between 0.1–0.3 deg^2. For the *wobble mode*, taken into account that the observation time is the same for ON and OFF, this value is simply $\alpha = 1/$(number of OFF regions), where the number of OFF regions can be 1 or 3 for standard observations.

The significance plot can also be given as a function of the alpha angle (see Sect. 2.4.3.3). Choosing this against θ^2 depends basically in the kind of analysis we want. The θ^2 plot is obtained for a source-independent analysis, since the angular distance is calculated with the *Disp method*, whose RF is trained with source-independent parameters, in the same way as the *hadronness*. On the other hand, the alpha plot is given for a source-dependent analysis where the position of the source is assumed, taken into account that the calculation of the alpha angle needs this information too. The alpha plots are commonly used for mono data analysis when the source coordinates are known. For the ON/OFF observations presented in this thesis (mono observations with MAGIC I), the alpha approach was used. Figure 2.31b shows an example of the alpha distribution.

Once the number of excess events is known, the significance can be computed roughly by *Significance* $= N_{exc}/\sqrt{N_{off}}$, a Gaussian approximation of Eq. 2.17 from Li and Ma (1983). Nevertheless, to report detection, Li & Ma significance is usually addressed:

$$\sigma_{\mathrm{LiMa}} = \sqrt{2\left(N_{\mathrm{on}}\ln\left[\frac{1+\alpha}{\alpha}\left(\frac{N_{\mathrm{on}}}{N_{\mathrm{on}}-N_{\mathrm{off}}}\right)\right] + N_{\mathrm{off}}\ln\left[(1+\alpha)\left(\frac{N_{\mathrm{off}}}{N_{\mathrm{on}}-N_{\mathrm{off}}}\right)\right]\right)}$$

(2.12)

2.4.3.10   Performance

Energy resolution and energy threshold The **energy resolution**, how accurately the instrument can determine the real energy of an event, is defined as the RMS of the Gaussian of all E_{bias} in each bin. Figure 2.32a shows the energy resolution and bias for pre- and post-upgraded periods and different Zd ranges. With the current system, the energy resolution reaches 15% at a few hundred of GeV, but gets worse at higher energies because of the higher probability of the image to lay at the edge of the camera (higher impact parameter) and get truncated. At lower energies, the

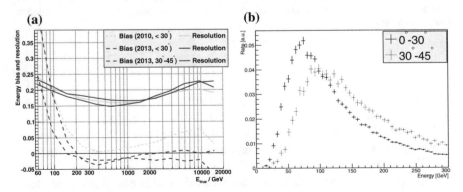

Fig. 2.32 *Left panel*: MAGIC energy bias (dashed lines) and energy resolution (solid lines) for the pre-ugrade (grey lines) and post-upgrade (red and blue lines) periods. *Right panel*: Rate of gamma rays after analysis cuts for low Zd angle (solid lines) and medium Zd angle (dashed lines). The peak of the distribution corresponds to the energy threshold, which is ∼75 GeV for Zd between 0–30° and ∼100 GeV between 30–45 °. Plots taken from Aleksić et al. (2016b)

energy resolution is as well worse, given the difficulty of reconstructing those images. During mono observations with MAGIC I, the energy resolution was 25% between 200 GeV and 1 TeV.

The **energy threshold** of the instrument is defined as the peak of the MC simulated energy distribution for a source that follows a power-law function with a standard photon index of $\Gamma = 2.6$. It is usually evaluated after analysis cuts, such as *hadron-ness*, *size* and θ^2 cuts, to obtain the distribution of surviving events. Figure 2.32b shows the energy distribution of simulated gamma rays at low and medium Zd range taken from Aleksić et al. (2016b). The current energy threshold of MAGIC (with a *size* cut of 50 phe) is ~75 GeV. Events with energy below the energy threshold can be detected too, but given the more complicated reconstruction of the image, the spectral points show large errors.

Sensitivity The sensitivity is defined as the minimum flux that the telescopes can detect in 50 h of observation with 5σ significance using the expression $Significance = N_{exc}/\sqrt{N_{off}}$. This flux is normally given in C.U., corresponding to a percentage of the Crab Nebula flux. For a certain observation with N_{exc} and N_{off} obtained for a time t, the significance at a time t_0 can be estimated as:

$$Significance(t_0) = \frac{N_{exc}}{\sqrt{N_{off}}}\sqrt{\frac{t_0}{t}} \qquad (2.13)$$

Equation 2.13 evidences the relation between significance and observation time, $\sigma \propto \sqrt{t_0}$. One can then compute the sensitivity in terms of Crab Nebula flux by assuming the standard definition where $t_0 = 50$ h and 5σ detection:

$$Sensitivity = \frac{5\sigma}{Significance(50)} \times C.U. \qquad (2.14)$$

In the same way as the flux, there are two ways to provide the sensitivity of an instrument: integral and differential sensitivity. The first one is obtained by applying the analysis cuts that give the best sensitivity above a certain energy threshold. In the case of MAGIC, the best integral sensitivity, $0.66 \pm 0.03\%$ C.U., is achieved above 220 GeV. Figure 2.33a shows the evolution of the MAGIC integral sensitivity as a function of the E_{thr} for different performances of the system. For the differential sensitivity, the cuts that provide the best sensitivity are searched for each energy bin. The pre- and post-upgraded MAGIC differential sensitivities are displayed in Fig. 2.33b.

Angular resolution Making use of the *Disp method*, the arrival direction of the gamma rays can be determined more accurately, which improves the **angular resolution**. If we consider a 2-dimensional distribution of the reconstructed arrival direction, the angular resolution (usually named as PSF) is commonly determined as the radius that contains 39% of the events. Nevertheless, in MAGIC the angular resolution is given for a 68% containment radius. Figure 2.34 presents the MAGIC angular resolution for stereo observations using DRS2 and DRS4. Currently, MAGIC obtains an angular resolution of ~0.10° at a few hundred GeV, reaching 0.06° in the TeV band.

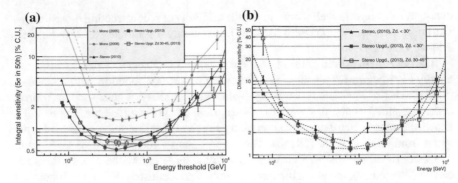

Fig. 2.33 MAGIC integral sensitivity as a function of the energy threshold (*left*) and differential sensitivity as a function of the energy (*right*). Integral sensitivity for stand-alone observations with MAGIC I are presented for the Siegen (light grey) and MUX (dark grey) readouts. Black line represents the sensitivity for stereoscopic observations before the upgrade period. The red and blue lines correspond to the current performance of the system for low and medium Zd, respectively. Plots from Aleksić et al. (2016b)

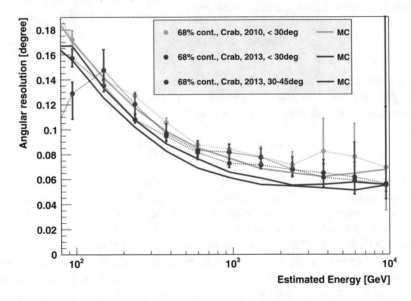

Fig. 2.34 MAGIC angular resolution for stereoscopic observations before upgrade (grey line) and after upgrade (red and blue lines). Plot taken from Aleksić et al. (2016b)

2.4.3.11   Skymaps

The skymaps are obtained by reconstructing the arrival direction of each gamma ray-like events, transforming the camera coordinates into sky coordinates (normally into RA and Dec), filling a 2-dimensional histogram with this information and subtracting the background. This task is performed by `Caspar` in the MARS framework. The

Fig. 2.35 TS skymap of the Crab Nebula

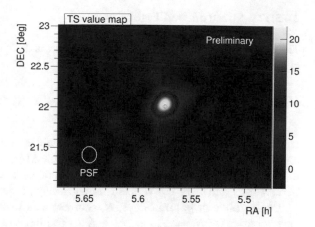

arrival direction is obtained by means of the *Disp method*, which prevent us to use source-dependent parameters and be biased in certain regions of the skymap.

Given the inhomogeneities of the camera, a key point when producing skymaps is to obtain reliable background histograms to subtract. This task becomes easy for the *wobble mode* observations: the background events are taken from the half of the camera where the source is not present. Thus, for two wobble positions, one would get two background histograms, one for each part of the camera. These two halves are then normalized by the time difference between them. Skymaps are usually given as relative excess histograms (by subtracting the background histograms to the gamma-like events histograms) or as Test Statistics (TS) skymaps. In both cases, the ON and OFF event histograms are smoothed for two reasons: first, a density function describes the events in each part of the camera (number of events in each point) given the camera inhomogeneity. This function is a Gaussian (Gaussian kernel) whose standard deviation can be varied to define the signal region (higher for extended sources). The skymap is also smeared according to the instrument PSF by a Gaussian distribution with standard deviation equal to the PSF, $\sigma_{Gauss} = PSF = 0.1°$. Tuhus, the total smoothing is:

$$\sigma_{smooth} = \sqrt{\sigma_{PSF}^2 + \sigma_{Kernel}^2} \tag{2.15}$$

For point-like sources, the σ_{Kernel} is usually equal to the PSF, and hence Eq. 2.15 can be simplified as:

$$\sigma_{smooth} = \sqrt{2} \cdot \sigma_{PSF} \tag{2.16}$$

An example of a skymap after smoothing is provided in Fig. 2.35.

2.4.3.12 Spectrum and Light Curve

The **differential gamma-ray energy spectrum** is defined as the number of gamma rays per unit area, time and energy that we observe. Mathematically, it is given by:

$$\frac{d\phi}{dE} = \frac{dN_\gamma(E)}{dEdA_{eff}(E)dt_{eff}} \quad [\text{photons TeV}^{-1}\text{cm}^{-2}\text{s}^{-1}] \quad (2.17)$$

where N_γ is the number of gamma rays in a certain energy range, A_{eff} is the so-called collection area and t_{eff} is the effective time. In the MAGIC framework, the computation of differential flux is performed by FLUX and Light Curve (`fluxlc`), for mono observations, and by FLUX versus Time and Energy (`flute`), for stereoscopic ones.

The **number of gamma rays**, N_γ, corresponds to the excess events within an energy range, i.e. $N_\gamma = N_{exc} = N_{on} - N_{off}$. The input to calculate this excess is the `Melibea` output: after all events passed through the RF trained branches and LUTs, each event has a defined *hadronness*, arrival direction and reconstructed energy. Thus, `Fluxlc` and `Flute` take the events and perform θ^2 (or alpha) and *hadronness* cuts for each energy bin to get the excess events. These cuts are commonly obtained by defining *efficiency* on the MC: we select a certain efficiency value for each variable, and the program changes the cut in each energy bin until the number of surviving events reaches the fixed efficiency. By default, the *hadronness efficiency* is set to 90%, while the θ^2 efficiency is fixed at 75%. These are looser cuts than those applied for the detection of the signal to guarantee a reliable collection area. The default *size* cut is set to 50 phe. Nevertheless, when analyzing data under the effect of the Moon, an increased *size* cut is used to remove events that can be produced NSB. This *size* cut is dependent on the moonlight level, as shown in Table 2.6.

Table 2.6 MAGIC *size* cuts for *moon* analysis

NSB ($\times$NSB$_{dark}$)	Equivalent DC [μA]	HV	*Size* cut [phe]
1–2	1.1–2.2	Standard	50
2–3	2.2–3.3	Standard	80
3–5	3.3–5.5	Standard	110
5–8	5.5–8.8	Standard	150
5–8	3.2–5.2	Reduced	135
8–12	5.2–7.8	Reduced	170
12–18	7.8 –11.6	Reduced	220
8–15	2.2–5.0	Standard with UV-filters	100
15–30	5.0–8.3	Standard with UV-filters	135

Table 2.7 Radius of the simulated collection area, A_{sim}, in MAGIC for different Zd ranges

Zd range [°]	A_{sim} radius [m]
5–35	350
35–50	500
50–62	700
62–70	1000

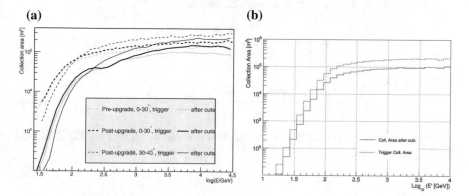

(a) **(b)**

Fig. 2.36 *Left panel*: Collection area before (dashed lines) and after (solid lines) analysis cuts for stereoscopic observations, taken from Aleksić et al. (2016b). *Right panel*: Typical collection area as a function of the energy for mono observations

The **collection area** is the geometrical area around the telescopes where the gamma rays are detected. It is calculated with MC events, applying to the simulated area, A_{sim}, a factor to account for the detection efficiency. A_{sim} is the collection area of an ideal instrument that would detect all simulated events for a given energy and Zd range, whose values according to Zd are listed in Table 2.7. Thus, A_{eff} is given as:

$$A_{eff}(E) = A_{sim} \frac{N_{surv}(E)}{N_{sim}(E)} \qquad (2.18)$$

where $N_{sim}(E)$ are the simulated events for a certain energy range and $N_{surv}(E)$ are the number of events that survive the analysis cuts (*hadronness*, θ^2, *size*) for a given energy range. Usually, MC gammas rays are simulated with a power-law function. In the case of MAGIC, the photon index is $\Gamma = 1.6$ (Fig. 2.36).

The last parameter to be calculated in order to compute the spectrum is the **effective time**. The effective time of a source is not identical to the elapsed time between the beginning and end of the observations, given the deadtime after storing each event and some gaps during data taking (e.g., between runs). The time difference between the arrival time of an event and the next one is Δt, which follows a Poissonian distribution with stable rate λ (without assuming deadtime):

$$P(n, t) = \frac{(\lambda t)^n}{n!} \cdot e^{-\lambda t} \qquad (2.19)$$

which is the probability of observing n events in a time t with a rate of λ. By definition, the probability that the consecutive event arrives after a time t is the same as observing 0 events before that time:

$$P(t_{next} > t) = P(0, t) = e^{-\lambda t} \tag{2.20}$$

The probability $P(t_{next} > t)$ is as well defined as:

$$P(t_{next} > t) = \int_t^\infty \frac{dP(t_{next} = t)}{dt} dt \tag{2.21}$$

and therefore:

$$\frac{dP(t_{next} = t)}{dt} = \lambda e^{-\lambda t} \tag{2.22}$$

Thus, the event rate is the product of the time evolution distribution (Eq. 2.22) and the stored events N_0:

$$\frac{dN}{dt} = N_0 \lambda e^{-\lambda t} \tag{2.23}$$

Now if we consider the deadtime (d), the fraction of lost events is $\lambda \cdot d$ and hence, the event rate would be:

$$\frac{dN}{dt} = N_{d,0} \lambda e^{-\lambda(t-d)} \tag{2.24}$$

Taken into account that the deadtime is fixed for a given observation, the distribution of the time differences of triggered events ($N_{d,0}$) is still exponential with a slope λ. We can get then the true rate of events, λ, by fitting an exponential to the distribution. The effective time is easily obtained dividing the number of triggered events by the true rate of events: $t_{eff} = N_{d,0}/\lambda$. Figure 2.37a shows a scheme of the time difference distribution for all and triggered events, while Fig. 2.37b presents a fit on real stereoscopic data with DRS4 (deadtime of 26 μs):

It is as well common to provide the Spectral Energy Distribution (SED), defined as:

$$E^2 \frac{d\phi}{dE} = E^2 \cdot \frac{dN_\gamma(E)}{dE dA_{eff}(E) dt_{eff}} \quad [\text{TeVcm}^{-2}\text{s}^{-1}] \tag{2.25}$$

which is normally used to study broadband spectra, since it shows the relative contribution of each wavelength to the total energy released by the source.

The flux can also be given in time intervals, the so-called **light curves**. In this case, the flux is integrated above a certain threshold E_0 in each time interval, i.e. it shows integral flux along time:

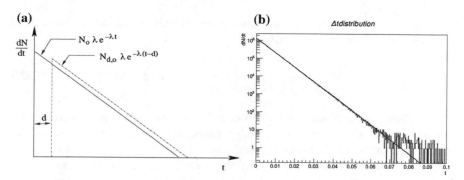

Fig. 2.37 *Left panel*: Scheme of the distribution of time differences for an ideal case in which all arrival events are observed (solid line) and taken into account a deadtime d (dashed line). *Right panel*: Distribution of time differences between triggered events. The fits provides the real rate events with which effective time is calculated

$$\phi_{E>E_0}(t) = \int_{E0}^{\infty} \frac{dN_\gamma(t)}{dEdA_{eff}(t)dt_{eff}(t)} dE \quad [\text{photons} \quad \text{cm}^{-2}\text{s}^{-1}] \tag{2.26}$$

For the computation of the light curve, the number of excess events is obtained from each time interval separately. In the same way, A_{eff} is as well get separately taking into account the Zd distribution in each interval.

2.4.3.13 Upper Limits

If no signal is found from the source, Upper limit (UL) on the flux are calculated. To compute the UL we make use, as before, of the excess events, N_{exc}, and the number of background events, N_{off}, and we assume a Confidence Level (C.L.) and a certain systematic error. Thus, we can calculate the N_{UL}, the maximum number of expected gamma-ray events, usually with the Rolke et al. (2005) method. In gamma-ray astronomy, it is common to use a 95% C.L. and particularly in MAGIC the systematic uncertainty assumed is 30% (see Sect. 2.4.3.14). The spectral shape of the source has to be defined as well. If no further information is available, a power-law function with photon index $\Gamma = 2.6$ (Crab-like spectrum) is usually assumed. The flux of the non-detected source is then defined as:

$$\phi(E) = K \cdot S(E) = K \cdot \left(\frac{E}{E_0}\right)^{-\Gamma} \tag{2.27}$$

and therefore, the integral flux above E_0 would be:

$$\int_{E_0}^{\infty} \phi(E)dE = K \int_{E_0}^{\infty} S(E)dE = \frac{N_{UL}}{\int_{E0}^{\infty} \int_0^{t_{eff}} A_{eff}(E)dEdt} \tag{2.28}$$

where t_{eff} is the effective time of the observation. The UL on the integral flux can be obtained easily from the above equation:

$$K_{UL} < \frac{N_{UL}}{t_{eff} \int_{E0}^{\infty} S(E)A_{eff}(E)dE} \quad \text{photons cm}^{-2}\text{s}^{-1}] \tag{2.29}$$

Full likelihood method The likelihood method is used on the estimation of model parameters that can describe a set of independent data. The conventional likelihood explores the existence of an astrophysical source based on Poissonian variables, i.e. number of detected events in the ON region (n) and number of detected events in the background region(s) (m). Thus, one can obtain the number of gamma rays (g) and background events (b) in the ON region by maximizing the likelihood function ($\mathcal{L}$):

$$\mathcal{L}(g, b|n, m) = \frac{(g+b)^n}{n!}e^{-(g+b)} \times \frac{(\tau b)^m}{m!}e^{-\tau b} \tag{2.30}$$

where τ is the normalization between the ON and the background region (which in *wobble mode* corresponds to the inverse number of *wobble* positions and in the ON/OFF method to the ratio of observational times in each case).

The full likelihood method increases the sensitivity by assuming the spectral shape of the source beforehand:

$$\mathcal{L}(g, b|n, m) = \frac{(g+b)^n}{n!}e^{-(g+b)} \times \frac{(\tau b)^m}{m!}e^{-\tau b} \times \prod_{i=1}^{n+m} \mathcal{P}(E_i) \tag{2.31}$$

where $\mathcal{P}$ is the Probability Density Function (PDF) of the event i with measured or estimated energy E_i.

This PDF is created with the hypothesis of the spectral shape and the Instrument Response Function (IRF) of each data sample. The IRFs are composed by the collection area, the effective time and the migration matrix (relation between E_{est} and E_{true}). The convolution of the spectral shape (usually a power law, $dN/dE \propto E^{-\Gamma}$) with the effective area and time allow us to obtain the number of expected events as a function of the true energy (E_{true}). This result, convolved in turn with the migration matrix, gives rise to the number of events as a function of the estimated energy. This is the value which allows to joint each independent data set (understood as those with different IRFs) into the likelihood expression.

The method tests the parameters of the assumed model that describes best our data, for which their values will change within a certain range until $\mathcal{L}$ is maximized. It is worth differentiating between the estimator, main parameter whose value we want to know, and nuisance parameters, which will be adjusted with the model but whose values are not important for the results we aim to obtain. For this thesis, a full likelihood method was not developed but instead we use the code of the method described in Aleksić et al. (2012a). The estimator in the original work was the cross-section of the interaction between two dark matter particles (for the annihilation scenario). In our case, this was simplified to the normalization factor of the spectrum, f_0. We applied

Fig. 2.38 Full likelihood distribution as a function of the estimated quantity, f_0

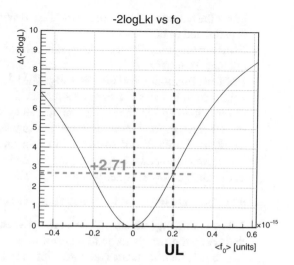

the full likelihood method in the study of Cygnus X-1, whose observations ranged from 2007 to 2014 covering almost entirely all MAGIC performances, leading this way to very different IRFs in multiple data samples.

On the other hand, it should be stressed that maximizing a variable is the same as minimizing its negative and maximizing a variable is equal to maximize its logarithm. This way, instead of maximizing $\mathcal{L}$, the method actually minimizes $-2\ln\mathcal{L}$. If $\mathcal{L}$ is the joint probability created from the product of PDF of independent samples, $-2\ln\mathcal{L}$ follows a χ-squared distribution with N degrees of freedom, which in our case is N = 1. Under this condition, the 95% C.L. value of the estimator is given by the variation of the $-2\ln\mathcal{L} + 2.71$ (see Fig. 2.38)

2.4.3.14 Systematic Uncertainties

The measurement of the gamma-ray spectra and light curves are affected by **systematic uncertainties**, summarized below:

1. The following systematic errors affect the *estimation of the gamma-ray energy*:

   - **Fluctuations in the atmospheric conditions**: The atmospheric *MagicWinter* model used in the MC simulations does not account for nightly changes in humidity, temperature, cloudiness or calima (fine sand arriving from the Sahara Desert). There is also a $\sim$15% variation in the atmospheric transmission from Summer to Winter. In a run-to-run basis the uncertainty was estimated to be $\sim$11% (Aleksić et al. 2016b).
   - **Light losses**: Dust in the mirrors or a wrongly alignment at the beginning of the night (as well as a not accurate enough bending model) lead to light losses. This is estimated to add $\sim$10% on the systematic uncertainties. Dust can also

affect the entrance of light in the PMTs and Winston cones, whose uncertainty is estimated to be 5%.

- **F-factor method**: The method used to calibrate the signal produces a systematic uncertainty of 10%.
- **PMT performance and flat–fielding**: The PMTs have temperature-dependent gain, which induces 2% error. Moreover, there are uncertainties in their QE (around 4%) and in the collection efficiency of the first dynode (~5%). On the other hand, the flat-fielding method (a procedure performed time to time to adjust the PMT HV to make the gain uniform) is performed with only one wavelength and it is temperature-dependent as well. This produces 6–8% uncertainty at low energies and less than 2% at energies greater than 300 GeV.

2. Below, there are listed the systematics that affects *the flux level*.

- **MC and data agreement**: Difference between the MC simulated gamma rays and real gamma rays leads to an error of the γ/hadron separation efficiency, which in turn, affects the calculation of the collection area.
- **Background estimation**: Dead pixels, dispersion on the PMT response or stars in the FoV can induce an error of 10–15%. This systematic error is reduced to ~1% when using *wobble pointing mode* for the post-upgrade period and to ~2% before this. On the other side, the background estimation uncertainty is larger as lower the energy is, so it affects mostly low-energy dominated sources, where the difference between N_{on} and N_{off} (signal-to-background ratio) is not that accentuated. In the latter case, the systematic can reach ~20% and, given that is energy-dependent, it can also affect the spectral index.
- **Telescopes mispoiting**: If telescopes point to different positions, the computation of stereo parameters used in the analysis is less reliable. The typical mispointing is less than 0.02° which leads to a systematic uncertainty on the gamma-ray efficiency around 4% (Aleksić et al. 2016b).
- **Higher NSB levels**: The higher the NSB, the more difficult to determine the image parameters due to fluctuations in the images. This can produce < 4% systematic errors.

To account for all these effects, an average 30% of systematic uncertainty is included in the Rolke et al. (2005) method to compute flux or ULs.

2.5   CTA

CTA is the future generation of ground-based observatory to study the VHE gamma rays within an energy range between 20 GeV and 300 TeV. It will improve the performance of the current IACTs abruptly: this next generation is expected to reach an angular resolution of <0.1° for most of the energies, with an energy resolution of 10–15%, a wide FoV of 8–10° and a ten times better sensitivity (see Fig. 2.40).

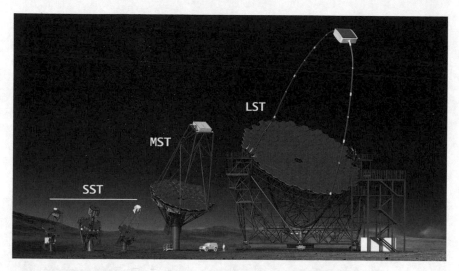

Fig. 2.39 CTA prototypes designs. *From left to right*: Three different SST, MST and LST. Picture taken from https://www.cta-observatory.org

All these improvements, along with the huge energy range covered, will allow us to expand our knowledge on several scientific subjects, which can be encompassed into three categories:

1. Study the origin of CRs: We aim to delve into how and where particles are accelerated and how they propagate within our Universe. Moreover, we want to study the impact on the surrounding medium of this CR propagation.
2. Explore extreme particle acceleration: CTA will allow us to probe extreme environments, to study in detail processes happening close to black holes, within relativistic jets or winds. Thus, we will understand better the mechanisms working behind different sources such as pulsars, plerions or microquasars.
3. Study the physics frontiers: with such low energy threshold, our goal is to shed light on the nature of dark matter and its distribution and get deeper into physics beyond the Standard Model.

For further information on the science planed with CTA, I refer the reader to Acharya et al. (2013).

CTA is planned to be formed by more than 100 IACTs distributed between the Northern and Southern hemispheres. The Northern array will be located in La Palma, at El Roque de los Muchachos along with the MAGIC telescopes. The Southern one will be placed in the Paranal Observatory (Chile). The proposed array layouts are presented in Fig. 2.41. The telescopes are divided into three types (see Fig. 2.39), which in turn correspond to different working sub-groups within the Collaboration:

- **LST**: The LST sub-group consists on more than 100 scientist from eight countries (Brazil, France, Germany, India, Italy, Japan, Spain and Sweden). These telescopes

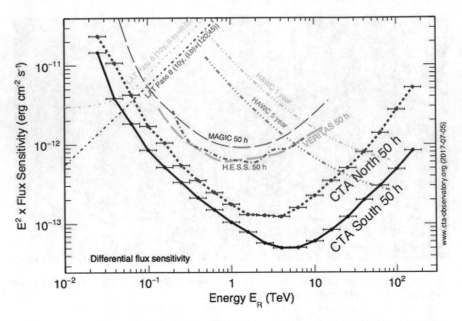

Fig. 2.40 Differential energy flux sensitivity of CTA compared to the existing gamma-ray instruments. The sensitivity curves for the Northern (blue points) and Southern (black points) hemisphere arrays are shown separately for 50 h of observations. Taken from https://www.cta-observatory.org

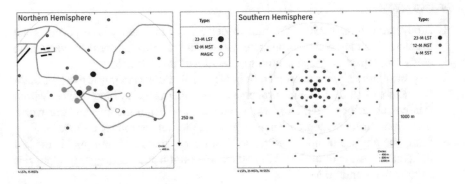

Fig. 2.41 Planned CTA layouts, taken from https://www.cta-observatory.org

are designed to achieve the lowest energies, down to 20 GeV. The planned baseline for CTA includes 4 LST in each hemisphere. They have a parabolic reflector of 23 m diameter held by a tubular structure made of carbon fiber and steel tubes. The light-structure, despite its 45 m height, allows the telescope to re-point within 20 s. With this fast re-positioning and low energy threshold, LST are thought to delve into galactic transients, GRB, high redshift Active Galactic Nuclei (AGN) studies and, in general, low-energy dominated sources. The LST camera weight less than two tons and it is composed by 1855 0.1° FoV PMTs, grouped in 265 clusters, that provides a total FoV of 4.5°. The PMTs present a QE of 42% and are equipped with light concentrators. The readout is performed by DRS4 chips. Deeper information regarding the LST prototype camera can be found in Chapt. 10 of this thesis. This prototype is expected to be fully installed in La Palma by mid-2018. During commissioning, structure and camera will be evaluated to modify the design of the next LST, if needed. After the commissioning phase, the prototype will be part of the Northern hemisphere array.

- **Medium Size Telescope (MST)**: The MST collaboration is formed by scientist from six countries: Brazil, France, Germany, Poland, Switzerland and the United States. The best sensitivity is achieved in an energy range between 100 GeV and 10 TeV, therefore they are designed to cover mid-energies. In total, CTA will host 40 MST: 25 in the Southern hemisphere and 15 in the Northern. The telescopes have a modified Davies-Cotton reflector of 12 m diameter with 16 m focal, held by a polar mount. Two pixelized cameras are designed for the MST: NectarCAM and FlashCAM. The former shares many characteristics with the one used in the LST. Currently, a MST prototype (without the final camera and readout) is installed in Berlin for testing purposes.
- **Small Size Telescope (SST)**: They are designed to cover an energy range between few TeV and 300 TeV, increasing CTA sensitivity at the highest energies. A total of 70 SST will be installed only in the Southern hemisphere. Three different prototypes were tested: one single-mirror design (SST-1M, with the collaboration of Poland and Switzerland) and two dual-mirror designs (SST-2M ASTRI, in which Brazil, Italy and South Africa are involved, and SST-2M GCT, with Australia, France, Germany, Japan, the Netherlands, the United Kingdom and the United States). All of them have 4 m diameter reflector dish and around 9° FoV.

2.6 Other Detectors

2.6.1 HAWC

The HAWC Observatory is the second generation of ground-based detectors which applies the Water Cherenkov technique to study the TeV gamma-ray regime (see Sect. 1.2.4). It is located in Sierra Negra, Mexico (19.0° N, 97.3° W, 4100 m a.s.l.), and it is the successor to the Milagro gamma-ray Observatory. The current system, inaugurated in March 2015, is comprised of 300 Water Cherenkov Detector (WCD) over an area of 22,000 m^2. Nevertheless, science operation started in August 2013, with a configuration of approximately one-third of the array (111 tanks; HAWC-111). HAWC operates with >95% duty cycle with a large instantaneous FoV of 15% of the sky, which allows it to scan two-thirds of sky every 24 h. Large effective area and duty cycle converts HAWC in an optimal instrument to perform survey studies on TeV sources. The dimensions of each tank are 7.3 m diameter and 5 m in height and attached to the bottom there are four PMTs: one high-QE 10-inch

Fig. 2.42 HAWC Observatory. http://www.hawc-observatory.org/

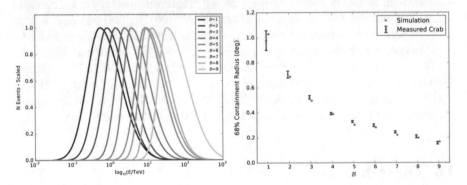

Fig. 2.43 *Left panel*: HAWC true energy distribution, split into 9 fractional hit bins (bin 1 corresponds to the ratio with less PMTs hit and bin 9 with the most). The plot is computed for a source with a power-law spectrum of index −2.63 and at a declination angle of +20°. At high Zd angles the image reconstruction becomes more difficult and the energy resolution worsens, leading to a broadening of the peaks. *Right panel*: HAWC angular resolution (for the 68% containment) as a function of each 9 bins. Plots taken from Abeysekara et al. (2017a)

Hamamatsu R7081-MOD placed in the center and three 8-inch Hamamatsu R5912 PMTs. Given its high altitude, HAWC can achieve the largest gamma-ray energies, from ∼100 GeV to ∼100 TeV. However, lower energy photons, although they can be detected, are much more difficult to discriminate from the background events. In this section, I will briefly comment on the energy, angular resolution and sensitivity of the detector. Nevertheless, the performance of the instrument is deeply presented in Abeysekara et al. (2017a), and the reader is referred to that work for further information (Fig. 2.42).

The data is divided according to the size of the event (ratio between the number of PMTs hit and the total number of PMTs operating) in 9 bins, where bin 1 implies less PMTs with signal and 9 corresponds to the maximum (see relation between energy bins and true energy in Fig. 2.43a). Given the PMT signal and arrival time, the core and direction of the cascade for each event are computed, whose errors increase at lower energies and large impact parameter from the detector. A good core fitter is crucial in HAWC to obtain, not only the reconstructed direction of the cascade, but also a good γ/hadron separation. Hadronic cascades produce a signal more randomly distributed all over the detector than EM showers: while hadronic showers give rise to clumpier and less compact images in the array, the gamma ray-induced showers are homogeneity and compact around the core position. HAWC makes use of this information to discriminate the nature of the EAS through two parameters, known as *compactness* and *PINCness* of the event. The *compactness* is the search of high signals outside of a 40 m radius area centered at the core of the cascade. Such signals far away from the core position reveal the presence of muons and hence, the hadronic origin of the shower. On the other hand, the *PINCness* is a kind of χ^2 fit to the lateral distribution of the cascade. A good fit is expected in case of EM showers, while hadronic lead to worse fits.

During my research sojourn in the Michigan Technological University, I could briefly work on the check of different options for the core fitter method. The calculation of the core position was performed by means of the χ^2 minimization, whose expression was defined as:

$$\chi^2 = \sum_{n}^{n=totalPMTs} \frac{(MPE - EPE)^2}{\sigma^2} \qquad (2.32)$$

where MPE is the measured number of phe get from the data, EPE the expected phe obtained from MC simulations assuming a certain lateral shower distribution and σ the deviation. The assumed lateral distribution needs to describes the density of charged particles as a function of the shower axis as accurate as possible. Thus, in HAWC, two assumptions were tested by that time: First of all, it was checked a simple Gaussian distribution, which was discarded for being not precise enough. Secondly, an Nishimura-Kamata-Greisen (NKG) distribution, described as follows:

$$EPE = \frac{N}{2\pi R_{Mol}^2} \frac{\Gamma(4.5 - s)}{\Gamma(4.5 - 2s)} \left(\frac{r}{R_{Mol}}\right)^{s-2} \left(1 + \frac{r}{R_{Mol}}\right)^{s-4.5} \qquad (2.33)$$

where N is the amplitude, R_{Mol}=124.21 m is the Molière radius in the air, r is the distance to the shower axis and s is the shower age.

The χ^2 was performed from $n = 0$ up to the maximum number of PMT operating on the data run we checked each time, using as initial input of the core position the center of mass of the shower image in the array. The process would then stop when $d\chi^2/(dxdydsdN)=0$, i.e. when the minimization is achieved. For simplicity, we first tried $\sigma = \sqrt{MPE}$ and $\sigma = \sqrt{EPE}$ in Eq. 2.32. Nevertheless, none of those assump-

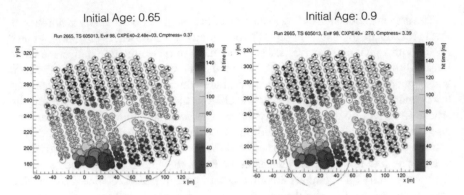

Fig. 2.44   2-D X core versus Y core plots in HAWC, where the core position obtained through χ^2 minimization method is marked with a red star. The initial age shower for the minimization is 0.65 (*left*) and 0.90 (*right*)

tions were good approximations: when assuming $\sigma = \sqrt{MPE}$ the code emphasized the small hits, while taking $\sigma = \sqrt{EPE}$ did so with larger hits. Therefore, the core position was not well defined. On the other hand, we found that the NKG distribution depended too much on initial parameters. As an example, Fig. 2.44 shows two different 2-D plots of the HAWC tanks where the shower images are highlighted and the core positions obtained through the χ^2 minimization method are marked:

For the plot on the left, the initial shower age was $s = 0.65$, while for the second was $s = 0.90$. Thus, it became obvious that this method depended strongly on the initial inputs and then it was not reliable. Therefore, some modifications were proposed: changes on the *EPE* distribution (described with an NKG/r distribution and the so-called *tank* distribution) and an introduction of a smearing into the σ parameter ($\sigma^2 = \sqrt{EPE + f^2 \cdot MPE}$). The *tank* distribution was referred to an NKG/r distribution with an average of the charge in each tank, which was much faster computationally. Nevertheless, information was lost in this process. In this case, a radial cut was included: the tanks inside a certain radius were not used to compute the core position as the signal there could be saturated, moving the core position and complicating this way the minimization. In order to prove all these changes, we compared proton and gamma-ray MC simulation with data making use of 2-D plots of X core against Y core (as the ones shown in Fig. 2.44) and the X core and Y core distributions. On the other hand, the precision of the algorithm was checked based on diverse parameters, as $\Delta Core = [(X - X_{MC})^2 + (Y - Y_{MC})^2]^{1/2}$. We conclude that the best option was to start doing a χ^2 minimization with the *tank* distribution, a radial cut of 10 m and a 35% of smearing to obtain the initial inputs for a second χ^2 minimization using individual PMTs with NKG/r distribution to get the final core position value. Nevertheless, the reader must note these were just preliminary tests and that this is not the final and current way HAWC calculates the core position, in which I did not work anymore during my thesis period.

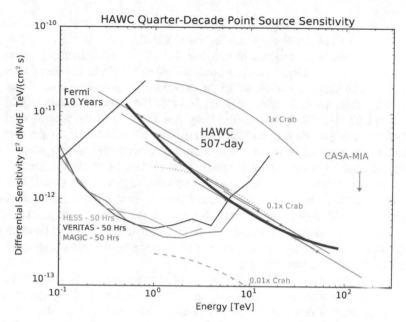

Fig. 2.45 HAWC sensitivity assuming a differential energy spectrum of $E^{-2.63}$. Observations of 507 days corresponds to approximately 3000 h. Taken from Abeysekara et al. (2017a)

The sensitivity of HAWC is shown in Fig. 2.45. Finally, a fit to the arrival time of the PMTs is performed in order to get the most probable shower front plane projection, which leads to the direction of the gamma ray and source. The angular resolution is therefore energy-dependent, as depicted in Fig. 2.43b, as more energetic cascades permit to compute a better fit. It ranges from 0.18° to 1.0° depending on the analysis bin.

2.6.2 Fermi-LAT

2.6.2.1 Performance and Analysis

The *Fermi* Gamma-ray Space Telescope (previously known as GLAST) is a satellite that studies the gamma-ray sky between 8 keV and ∼500 GeV. The satellite is formed by two instruments: Large Area Telescope (LAT) (∼20 MeV to ∼500 GeV), and GLAST Burst Monitor (GBM) (8 keV to ∼30 MeV), being the former the one from which I analyzed data in this thesis.

The LAT covers a 2 sr FoV, which allows it to observe 20% of the sky at any moment. It can observe in two modes, the sky-survey mode that covers the entire sky every three hours, and the pointing mode during which the satellite is observing a certain source. It is composed by four subsystems: tracker, calorimeter, anticoincidence detector and data acquisition system. When a particle enters in the LAT, the first interaction occurs in the anticoincidence detector. This device is in charge of rejecting hadronic events. Charged particles produce flashes of light in the anticoincidence detector, which sends a veto signal to the data acquisition system to avoid saving the event. The rejection efficiency reaches 99.97%. The software of data acquisition system in LAT also discriminates events based on the arrival direction to avoid, for example storing gamma rays coming from the Earth's atmosphere. If the primary particle is a gamma ray, it passes through the anticoincidence detector without any effect and interacts in one of the 16 thin tungsten sheets placed in the tracker. The interaction gives rise to an electron and positron due to pair production. Given that the energy of the gamma ray is much larger than the rest mass of the electron and positron, they will keep the same track direction as the primary gamma. Thus, the reconstruction arrival of the gamma ray can be computed. Finally, the electron and positron reach the calorimeter that provides their energies, obtaining in this way the energy of the incident gamma ray.

The response of *Fermi*-LAT to gamma rays of a certain energy and arrival direction (in instrument coordinates) is defined by the IRFs. Here, I will briefly comment on the Pass 8 Release 2 Version 6 (P8R2-V6) data, which was the one analyzed for this thesis. For the analysis, we select events from a certain *event class*, which includes a set of unique response functions. Formerly, the *event class* was just divided into two *event types*: FRONT and BACK events, accounting from the location of the tracker in which the pair production takes place. From the top of the instrument, the tracker device is formed by 12 layers with 3% radiation length converters (the so-called FRONT or thin section), followed by 4 layers of 18% radiation length (BACK or thick section). Multiple-scattering is more probable to happen in thicker regions and hence, the angular resolution is better (approximately by a factor of two) on FRONT photons. FRONT and BACK events present different IRFs. With the Pass 8 release, two new partitions of the event type were possible:

- **PSF event type**: It separates the data depending on the quality of the reconstruction direction (and hence PSF). Data is divided in four quartiles, where PSF_0 corresponds to the poorest quality and PSF_3 to the highest.
- **EDISP event type**: In the same way as for the PSF, the four quartiles in this case represents the quality of the energy reconstruction.

Each quartile has its own IRFs. In Fig. 2.46 the effective area, PSF, energy resolution and sensitivity as a function of energy is shown for FRONT and BACK events. For comparison, it is also displayed the result of using both FRONT+BACK.

Data format[3] provided by *Fermi*-LAT are event lists, to which selection criteria that best fit to a specific scientific purpose are applied. First of all, cuts based on position, time, energy range, RoI, event class or type or maximum Zd are computed to set up the data sample to high-level analysis. Some of these cuts are already recommended

---

[3]*Fermi*-LAT data is publicly available at the Science Support Center, https://fermi.gsfc.nasa.gov/cgi-bin/ssc/LAT/LATDataQuery.cgi.

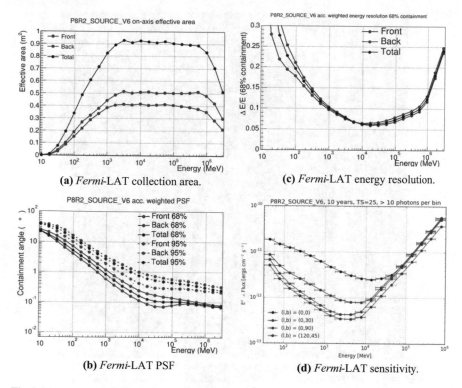

Fig. 2.46 *Fermi*-LAT IRFs and performance plots. Plots taken from https://www.slac.stanford.edu/exp/glast/groups/canda/lat_Performance.htm

by the *Fermi*-LAT Collaboration, while others are subjective and depend on each analysis. Among the recommended ones, it is worth highlighting the so-called in the *Fermi*-LAT nomenclature *evclass* = 128, which includes only those events with high probability of being photons. There is also a maximum cut in zenith angle, Zd=90° to avoid contamination from atmospheric gammas coming from the Earth's limb (that arrives from angles of ∼110°). The latter becomes more important for weak sources. This selection of events is performed by the program gtselect. Cuts on time intervals can also be applied. Usually they are computed to remove data from periods in which the spacecraft presented problems. The recommended cut to do so is DATA_QUAL> 0&& LAT_CONFIG==1 and is done by means of gtmktime.

After the data selection, one can obtain the counts map (CMAP) and exposure maps (the latter is needed for further steps of the analysis). The CMAPs represent the number of counts for each spatial bin, whose size (typically 0.2°bin) is defined by the analyzer.

The count map for the entire energy range (see Fig. 2.47) is useful to check any failed cut, as inconsistent structures could appear in the FoV in that case. The likelihood method, used later on in the *Fermi*-LAT analysis, needs a 3-D CMAP. This

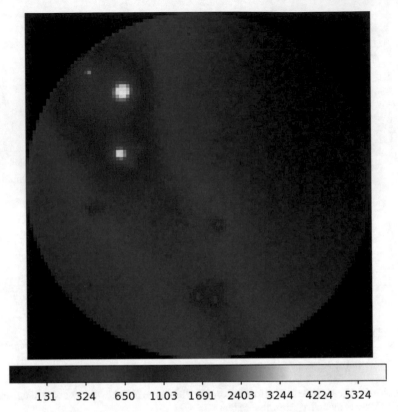

131   324   650   1103   1691   2403   3244   4224   5324

Fig. 2.47 Example of a *Fermi*-LAT CMAP (between 100 MeV and 500 GeV) centered at Cygnus X-1

is simply understood as a CMAP (counts VS space) in each energy range selected. These are obtained with the program gtbin. The energy bins applied here will affect directly the precision of the likelihood: if the bins are too big the IRFs will not be accurately defined in each range (e.g., at lower energies the PSF is worse than at higher). Normally, 10 bins per decade of energy are recommended.

The exposure maps give the amount of time that the instrument observed the region we aim to analyze. To obtain it, one needs to first compute the so-called livetime cubes, with the gtltcube tool, which are healpix grid[4] that provide the observation time for each source on the sky as a function of the inclination angle (between the normal line of the satellite and the position of the source). According to the livetime calculated, the gtexpcube2 tool generates a binned exposure map (same spacial and energy bins as before). The exposure map can be done for a certain region around our RoI (always larger to account for the contribution of other sources) or for the entire sky. The consuming time in the latter is just slightly higher.

---

[4]http://healpix.jpl.nasa.gov/.

On the other hand, the likelihood method need a source model (XML file) as an input. The source model is defined by the position of the target source and all sources including in the RoI, as well as their spectral shape and parameters. Usually, point-like background can be defined from the public 4-year point source catalog Third *Fermi*-LAT cataog (3FGL)[5] (which contains data in the 100 MeV to 300 GeV energy range) to create the model, while additional sources from any FHL (above 10 or 50 GeV) can be included depending on the energy range of the analysis. Taken into account that the available catalogs were published years ago before the release of the improved Pass 8 data, during the current *Fermi*-LAT analysis several point-like sources, that work as background for the analysis, arise in the residual maps. These sources have to be added manually to the XML file (see, e.g. Sect. 4.2.2). At low energies, *Fermi*-LAT presents a PSF as poor as $\sim$3.5° while at high energies the angular resolution is $\lesssim$0.15°Therefore, emission from the target source can be affected by that coming from the neighbors. Thus, spectral parameters of sources nearby the source of interest are left free, while the parameters of other objects located further away can be totally fixed (or let free only the normalization parameter) to accelerate the computation. Highly variable or very bright sources can be left free as well. Besides the point-like background, the model needs to account for the isotropic diffuse background model, i.e. gamma rays coming from cosmic-ray interaction. These models are provided by the *Fermi*-LAT Collaboration, which for the Pass 8 data correspond to the *gll_iem_v06* and *iso_P8R2_SOURCE_V6_v06* files for the galactic and extragalactic contribution, respectively.

At this point of the analysis, one can compute the so-called source map with the gtsrcmaps tools. This is a model counts map that will be applied in the likelihood method. The gtsrcmaps makes use of the spectrum for each source defined in the XML file, which is multiplied by the exposure of the position of the source in the sky (exposure map) and convolved with the effective PSF. Once this is ready, the likelihood method can be computed.

The likelihood is simply defined as the probability of obtaining the observed number of counts given a certain model. The model is the one provided as input (XML file). The likelihood tools need to optimize simultaneously the value of those parameters left free inside the model of several sources. The best parameter values and their uncertainties are those that maximize the likelihood expression. First of all, as shown along this section, one needs to take into account that each count is characterized by different observable, e.g. its energy, its inclination angle and its event type. This implies that the *Fermi*-LAT analysis is a multi-dimensional analysis (multiple number of bins), leading to small statistics in each bin. Therefore, the distribution in each bin is Poissonian. The likelihood can be then defined as the product of the probability of observing the detected counts in each bin:

- Being m_i the number of expected counts given a certain model in each bin i,
- the probability of detecting n_i counts in that bin can be defined by $p_i = m_i^{n_i} e^{-m_i}/n_i!$, and therefore

---

[5]https://fermi.gsfc.nasa.gov/ssc/data/access/lat/4yr_catalog/.

the likelihood, $\mathcal{L}$, is the product of all p_i in all i, which can be expressed as:

$$\mathcal{L} = \exp[-N_{exp}] \prod_i \frac{m_i^{n_i}}{n_i!} \tag{2.34}$$

where N_{exp} is the total number of expected counts given the assumed model. Therefore, N_{exp} and m_i depend on the model and n_i on the data. Equation 2.34 is the expression for the so-called binned *Fermi*-LAT analysis, and the aforementioned described steps correspond to this type of analysis, which is applied to large data samples. In this thesis, I analyzed sets of years of *Fermi*-LAT data and hence, this kind of analysis fitted my scientific purposes. The reader is encouraged to follow the *Fermi*-LAT tutorial available in the web page[6] for deeper information of both binned and unbinned (recommended for small data samples) analysis.

Usually, the maximum likelihood value is mapping out over a grid of coordinates, for which it is used the TS quantity. This parameter is defined as follows:

$$TS = -2ln \left(\frac{\mathcal{L}_{max,null}}{\mathcal{L}_{max,source}} \right) \tag{2.35}$$

where $\mathcal{L}_{null}$ and $\mathcal{L}_{max,source}$ are the maximum likelihood values without (*null hypothesis*) and with an additional target source in the model, respectively.

On the other hand, SED can be produced by performing the maximum likelihood analysis in each energy bin separately, while keeping the background spectral parameters fixed to those values obtained in an overall search. Normally, not all spectral parameters of the target source are left free either: for a power-law distribution, as it is the case of Cygnus X-1 (source discussed in this thesis), the flux normalization is left free to vary in each energy bin, whilst the spectral index is set to that obtained in the overall fit.

Finally, in order to obtain a lightcurve, I used a set of scripts to download data directly from the *Fermi*-LAT web page in a daily basis (accounting for those days in which the spacecraft was not in a pointing mode) and to perform the analysis based on data selection criteria and background model created beforehand. Due to the computational consumption of a day-to-day analysis over ~8 years of *Fermi*-LAT data, as that performed for this thesis in Cygnus X-1, only spectral parameters of highly variable sources are left free in the background model, besides the target source.

References

Abeysekara AU et al (2017a) ArXiv e-prints
Acharya BS et al (2013) Astropart. Phys. 43:3
Aharonian F et al (1997) Astropart. Phys. 6:343

---

[6]https://fermi.gsfc.nasa.gov/ssc/data/analysis/.

Albert J et al (2008) Nuclear Instrum. Methods Phys. Res. A 588:424

Aleksić J et al (2012a) J. Cosmol. Astropart. Phys. 10:032

Aleksić J et al (2012b) Astropart. Phys. 35:435

Aleksić J et al (2016a) Astropart. Phys. 72:61

Aleksić J et al (2016b) Astropart. Phys. 72:76

Aliu E et al (2009b) Astropart. Phys. 30:293

Aliu E et al (2009a) Astropart. Phys. 30:293

Bitossi M (2009) Ultra-fast sampling and readout for the MAGIC-II telescope data acquisition system, PhD thesis

Bretz T et al (2009) Astropart. Phys. 31:92

Cherenkov P (1934) C. R. (Doklady) Akad. Sci. URSS 2:451

Fomin VP et al (1994) Astropart. Phys. 2:137

Fruck C et al (2014) arXiv:1403.3591

Gini C (1921) Econ. J. 31:22

Hofmann W et al (1999) Astropart. Phys. 12:135

Lessard RW et al (2001) Astropart. Phys. 15:1

Li T-P et al (1983) Astrophys. J. 272:317

Lombardi S (2011) Int. Cosmic Ray Conf. 3:266

López-Coto R (2015) Very-high-energy γ-ray observations of pulsar wind nebulae and catacysmic variable stars with MAGIC and development of trigger systems for IACTs, PhD thesis

MAGIC Collaboration et al (2017) ArXiv e-prints

Majumdar P et al (2005) Int. Cosmic Ray Conf. 5:203

Mazin M (2007) A study of very high energy gamma-ray emission from AGNs and constraints on the extragalactic background light, PhD thesis

Mirzoyan R (1997) Int. Cosmic Ray Conf. 7:265

Nakajima D et al (2013) Proceedings of the ICRC 2013, id 787

Rolke WA et al (2005) Nuclear Instrum. Methods Phys. Res. A 551:493

Sitarek J et al (2013) Nucl. Instrum. Methods Phys. Res. A 723:109

Tescaro D et al (2013) arXiv:1310.1565

Wagner R (2006) Measurement of VHE γ-ray emission from four blazars using the MAGIC telescope and a comparative blazar study, PhD thesis

Zanin R (2011) Observation of the Crab pulsar wind nebula and micro quasar candidates with MAGIC, PhD thesis

Zanin R et al (2013) in Proceedings of the 33st international cosmic ray conference, Rio de Janeiro, Brasil

Part II
Microquasars in the Very High-Energy Gamma-Ray Regime

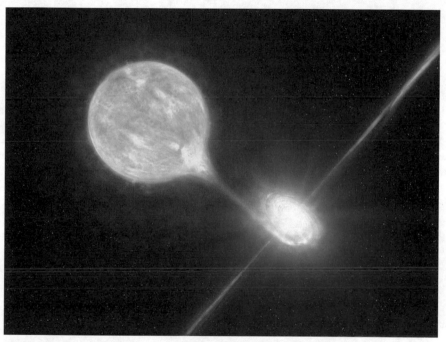

Fig. II.1 Artist's impression of a microquasar. Credit: ESO/L. Calçada

Chapter 3
Microquasars, Binary Systems with Powerful Jets

3.1 X-Ray Binaries

Our Galaxy contains hundred billions of stars with very different features, masses, size and ages. The understanding of their birth, evolution and death was, and still remains to be, a goal to achieve for the astronomers. The life of the stars begins within molecular clouds, where gravity is responsible of joining the dust and gas of the environment to give rise to the celestial objects. If the created body has enough mass to reach high temperatures and fuse Hydrogen (H) atoms into He in its core, then a main sequence star is born. Most of the stars in the Milky Way belong to this type. The main sequence stars are constantly producing energy via nuclear fusion which counteracts the pressure caused by gravity. The fate that a main sequence star follows depends strongly on its initial mass and composition (Heger et al. 2003).

For small stars ($M_\star$ < 9-10 $M_\odot$), the mass is not enough to increase the temperature to reach the Carbon (C) fusion point. When this happens, the nuclear fusion ceases and the gravitational pressure starts to dominate. However, as the matter is compressed, the electrons that form this mass get closer, which need to stay in different energy levels given the Pauli exclusion principle. This results into an electron degeneracy that causes internal pressure against the collapse. This degeneracy state is only achieved if the mass of the star, after the fusion processes stop, is lower than the Chandrasekhar limit of 1.4 $M_\odot$ (Chandrasekhar 1931), generating a remnant known as White Dwarf (WD).

In some cases, the core of the star presents layers which reach the necessary temperature to fuse He atoms into C in their inward collapse. The pressure originated during this fusion expands the star enlarging its size until forming a red giant. Nevertheless, this is just a temporary phase before the inevitable collapse happens. In the case of more massive stars (9-10 $M_\odot$ < $M_\star$ < 40 $M_\odot$), the temperature is high enough and fusion of heavy elements like C, Neon (Ne), Oxygen(O), Silicon(Si) or Iron(Fe) takes place. If the mass of the star at the end of its productive phase overpasses the Chandrasekhar mass, the gravitational pressure cannot be halted by

© Springer Nature Switzerland AG 2018
A. Fernández Barral, *Extreme Particle Acceleration in Microquasar Jets and Pulsar Wind Nebulae with the MAGIC Telescopes*, Springer Theses, https://doi.org/10.1007/978-3-319-97538-2_3

electron degeneracy: the star collapses giving rise to a huge thermonuclear explosion, the so-called SN. The different types of SNe will be discussed more deeply in Appendix A. The remnant behind such an explosion can be a NS or a BH. The former is created when the mass of the star ranges from 9-10 $M_\odot$ to 25 $M_\odot$. Once the electron degeneracy pressure is overcome, the core keeps collapsing while its temperature increases. At this point, electrons and protons combine via electron capture, which produces neutrons and neutrinos. When the core reaches a density of the order of the nuclear one, the neutron degeneracy stops the collapse in the same way electron degeneracy did before, and the infalling outer layers are expelled in the SN, leaving a NS as remnant. The BH is created when the mass of the star is higher ($25\,M_\odot < M_\star < 40\,M_\odot$) and the gravitational pressure dominates over the neutron degeneracy. At even higher mass ranges ($40\,M_\odot < M_\star < 140\,M_\odot$ or $M_\star > 260$ $M_\odot$), the BH is formed directly with no visible SN. For a particular case between 140 $M_\odot$ and 260 $M_\odot$, a so-called pair-instability SN takes place without leaving any remnant after the explosion.

When these compact objects (NS or BH) orbit and accrete material from an optical star (usually a main-sequence one), an **X-ray binary system** is formed. These kind of sources are extremely luminous in the X-ray band ($L_X \sim 10^{35} - 10^{38}$ erg s^{-1}) and the brightest compact sources in the medium X-ray regime, from 2 to 10 keV (Grimm et al. 2002). Given their luminosity, X-ray binaries were the main subject during the first years of X-ray astronomy, until the 1980s, when X-ray imaging instruments (first Uhuru, Ariel 5, HEAO-1 and later, EXOSAT, Ginga, RXTE, ROSAT and BeppoSAX, among others) opened the door to fainter sources. Thus, from the first non-solar X-ray source discovered in 1962, Scorpius X-1 (Giacconi et al. 1962) which afterward was confirmed as a binary system (Gursky 1966), up to the present, large number of satellite missions allowed us to detect $\sim$200 X-ray binaries and to extend our knowledge on the physical properties of these sources.

As explained before, the death of a star can end up in a WD as well. It is, therefore, worth mentioning that if the compact object of the binary is a WD, instead of a BH or NS, then the system is named Cataclysmic Variable (CV). Although WDs are very dense, the efficiency converting gravitational energy into X-rays is small (around 0.03%) compared to systems harboring NSs or BHs ($\sim$10% and 40%, respectively). Therefore, most of the transferred matter is released in the optical or UV bands, where only a minor fraction is released in X-rays. Consequently, these sources are not usually cataloged as X-ray binaries.

One of the most established ways to classify X-ray binaries is according to the nature of the main sequence or companion star (a.k.a. donor or secondary star). Thus, the X-ray binary systems can be split into two groups: High-Mass X-ray Binaries (HMXBs), if $M_{donor} \gtrsim 10\,M_\odot$, and Low-Mass X-ray Binaries (LMXBs), if $M_{donor} \lesssim 2\,M_\odot$. The dominant mass-transfer process in these two types of binaries is different (stellar wind-driven for the HMXBs and Roche-lobe overflow in LMXBs, as explained later), which allows to solve the unclear classification for X-ray binaries with 2–10 $M_\odot$.

3.1.1 High-Mass X-Ray Binary Systems

In the HMXBs, the primary star, from whose death the compact object arises, was previously forming a star binary system. Since this star was more massive than its companion, evolved faster, becoming a giant and loosing its outer layers onto the companion, until it finally died in a SN. If at the end of its life, its mass is lower than the companion's, the binary system remains together, otherwise, it will be disrupted.

These systems are comprised of a massive secondary early O or B-type companion star, which presents strong stellar winds, with a mass-loss rate around $10^{-6} - 10^{-10}$ $M_\odot$ yr^{-1}, and a terminal velocity up to 2000 km s^{-1}. This type of stars displays high optical, UV and IR luminosity, which dominates the total emission from the system ($L_{optical}/L_X > 1$). The accretion on these sources is produced in a high-velocity wind-driven form, where the compact object captures a fraction of the OB star wind that passes within a certain radius, called *capture* or *accretion radius*, below which the matter cannot avoid the gravitational attraction of the BH or NS (see e.g. Bondi 1952).

Additionally to this dominant mass transfer, a secondary transfer through the so-called process Roche-lobe overflow can occur. This type of accretion takes place when the donor star fills its Roche lobe, region in which the material is gravitationally bound to the star. The material that overpasses this lobe will fall onto the compact object via the first Lagrangian point, where the gravitational forces from the BH or NS and the donor star are equal (see Fig. 3.1). However, if the compact object's mass is greater than the companion's, the Roche-lobe overflow will become unstable after $\sim 10^5$ yr from its beginning (Savonije 1983).

The accreted material presents angular momentum and the conservation of this magnitude prevents the matter from falling directly onto the compact object. Thus, the mass transfer is produced through an accretion disk formed around the compact object (see Fig. 3.2), which is the most common mode of accretion in astrophysics. The heating created by friction inside the accretion disk produces the X-rays, which give the general name to these systems. In the case of HMXBs, the X-ray emission peaks at $kT \geqslant 15$ keV, characterized normally by regular X-ray flux variations (Camenzind 2007). Among other modulations, X-ray binaries normally suffer flux variation in an orbital period scale, considered as the time that the compact object needs to complete an orbit around its companion. This modulation can be seen in different wavelengths, from the optical to higher energies, and is produced by the absorption or scattering of the radiation by photons from the companion star or accretion disk. HMXBs present orbital periods from days to hundreds of days.

The lifetime of HMXBs is short, around $10^5 - 10^7$ yr, due to the high-mass of their secondary stars. This makes the HMXBs distribute along the galactic plane among the young stellar populations (see Fig. 3.3).

The HMXBs can be, in turn, classified into two principal sub-groups according to type of the secondary star: Be/X-ray and Supergiant X-ray binary systems. A small fraction of the HMXBs, around the 18%, do not belong to any of these two

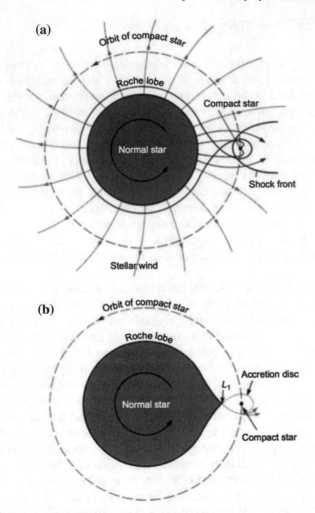

Fig. 3.1 Sketches that illustrate the two different accretion processes in binary systems. In **a**, the compact object is embedded in the strong stellar wind from its massive companion star, giving rise to the wind-driven mass transfer. In **b**, the companion star expands filling its Roche lobe and the material falls onto the compact object through the Lagrangian point, L_1. In both cases, the matter does not reach the compact object directly but through an accretion disk (Longair 2011)

sub-classes. In this group, it would be fit the X Per like systems (as suggested by Reig et al. 1999) or systems harboring Wolf Rayet (WR) stars.

3.1.1.1   Be/X-Ray Binaries

The Be/X-ray binaries sub-class represent the largest population of HMXBs with a 57%. These very luminous systems ($L_X \sim 10^{36} - 10^{38}$ erg s^{-1}) are composed of a Be companion star surrounded by a circumstellar envelope with disk-like geometry

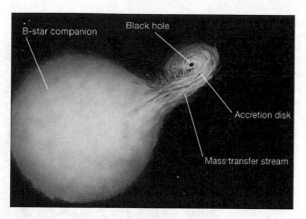

Fig. 3.2 Artist's vision of an X-ray binary composed of a BH and a massive star. Credit: L. Chaisson

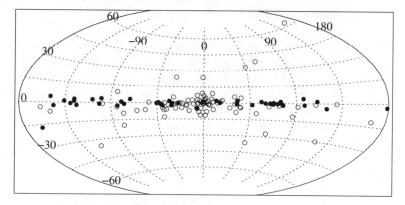

Fig. 3.3 Distribution of 52 HMXBs (black circles) and 86 LMXBs (empty circles) in galactic coordinates. HMXBs are distributed along the galactic plane and LMXBs are concentrated in the galactic center and globular clusters (Grimm et al. 2002)

(Fig. 3.4), which displays Balmer line emission and causes an excess in the IR band. The physical formation of this circumstellar disk, from which the accretion takes place, is not well understood yet, although it is thought to be related with the fast rotation of the star. The compact object is normally a NS (or pulsar in case of rapid rotation and high magnetic field, see Sect. 7.1) and its passage through the circumstellar disk of the donor is the responsible of the X-ray emission. These sources also display X-ray outbursts that can be divided into two types: periodic outbursts at the periastron of the orbit (nearest point to the companion star), classified as *Type I*, and non-regular huge outbursts or *Type II* outbursts, produced by an expansion of the circumstellar disk. This sub-group of HMXBs usually follows a highly eccentric orbit, which intensifies *Type I* outburts. For an extended description of the systems see Camenzind (2007).

Fig. 3.4 Be/X-ray binary composed of a pulsar, whose interaction with the circumstellar disk of the Be star gives rise to X-ray emission and even higher energies as gamma rays (Mirabel 2006)

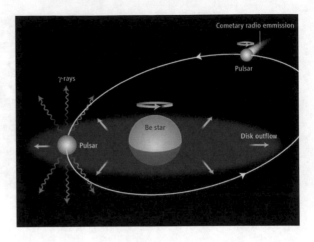

3.1.1.2   Supergiant X-Ray Binaries

The supergiant X-ray binary systems are also composed of a secondary OB type star or even type A. The main difference with respect to the Be/X-ray binaries is the mass-transfer process. Due to the lack of a circumstellar disk in these sources, the accretion arises from stellar wind outflow. In this case, the compact object (either a BH or a NS) follows normally a circular orbit around the companion. The steady wind outflow prevents the system to suffer as many X-ray outbursts as in Be/X-ray binaries and produces, in turn, persistent and less luminous X-ray emission ($L_X \sim 10^{34} - 10^{35}$ erg s^{-1}). The percentage of HMXBs that belong to this type reaches the 25%.

3.1.2   Low-Mass X-Ray Binary Systems

The origin of LMXBs is less clear, but they are thought to be created by capture: the compact object, formed after the SN explosion, interacts with a close cluster and captures a low-mass star due to its strong gravitational force.

These systems are composed of donor stars with type later than A, i.e. low-mass G, K or M companions. Given the faint secondary stars ($L_{optical}/L_X \ll 1$), the optical spectrum is dominated by radiation originated at the accretion disk by reprocessed X-ray emission. Actually, the contribution of the donor is only not negligible in cases in which the LMXB arises from an intermediate binary. The late type stars forming the LMXBs do not have strong wind to produce accretion via wind-driven process. This way, the system is powered by the Roche-lobe overflow (Fig. 3.1). The X-ray spectrum in these binaries is softer than in their counterparts, peaking at $kT \leq 15$ keV. LMXBs also display weaker magnetic fields ($\sim 10^9 - 10^{11}$ G) than HMXBs ($\sim 10^{12}$ G), which allow X-ray bursts to happen (produced by thermonuclear fusion of accreted material in the surface of the compact object, Camenzind 2007).

LMXBs are more compact than the HMXBs, i.e. the separation between the star and its accretor is smaller. While the size of the LMXBs is generally around a solar radius, the size of the HMXBs can reach tens of solar radii. Because of this compactness, their orbital period is lower as well, ranging from minutes to hours (see e.g. Table 1.1 of Lewin et al. 1995). On the other side, the lifetime of these systems is higher than the HMXBs', between 10^7 and 10^9 yr, determined mostly by the accretion process. As in the case of old stellar populations, LMXBs are mostly located close to the galactic center and in globular clusters (see Fig. 3.3).

There is a sub-class of X-ray binaries, in both HMXBs and LMXBs, characterized by the presence of extended radio-emitting jets. These systems are known as **Microquasars** (Mirabel et al. 1999) and constitute one of the main topics of this thesis.

3.2 Microquasars

Microquasars are a sub-type of X-ray binary systems. Therefore, they are composed of the same elements presented in Sect. 3.1: a high or low-mass companion star and a compact object (either BH or NS) that accretes material through an accretion disk of size $\sim 10^3$ km. However, normally two-sided highly collimated ($<15°$ opening angle, as defined by Mirabel et al. 1999) streams of fluid, gas or plasma, the so-called jets, are launched perpendicularly from the compact object (see Fig. 3.5).

These jets, whose speed is usually characterized by the Lorentz factor,[1] are very powerful non-thermal emitters detected at different wavelengths. Nevertheless, they are not steady structures: their presence or absence seems to depend on the accretion rate, which allows to distinguish different X-ray states, as it will be deeply discussed in Sect. 3.2.3.1.

The name of microquasar, chosen by Mirabel et al. (1992), born from the analogy between these objects and their scaled-up counterparts, the quasars. The latter are distant AGN from which the first evidence of jet-like structures was obtained. These jets were discovered in the optical regime by Curtis (1918), emanating from the galaxy M87. In quasars, this relativistic ejecta can travel several million parsecs, well above the distance reached by microquasars (around few parsecs). The morphological analogy does not only concern the existence of relativistic outflows: quasars also host BH as central objects which accrete material from the surrounding. The main difference is the scale of the BH in each case. While the microquasars present stellar-mass BHs, the ones forming quasars are super-massive BHs of the order of several million solar masses (Rees 1998). On the other hand, the accretion disk of the quasars (with size $\sim 10^9$ km) is not fed from a companion stellar object but from the ISM of their galaxies and from disrupted stars due to their strong gravitational force.

---

[1] Lorentz factor described as $\Gamma = (1 - v^2/c^2)^{-1/2}$, where v is the flow velocity and c the speed of light. In BH microquasars, the Lorentz factor can reach $\Gamma \sim 2.5$ ($0.92c$).

Fig. 3.5 Anatomy of a
microquasar where the
different elements that
composed the system are
labeled. Credit: Imago
Mundi

Nevertheless, the thermal temperatures achieved in the quasars' accretion disks by viscous/friction dissipation is of the order of several thousand degrees, instead of several million degrees as in the case of the microquasars. The reason is that as more massive the BH is, the cooler will be its accretion disk. Given by Rees (1984), the characteristic blackbody temperature at the last stable orbit of the disk for a BH accreting at the Eddington limit, defined in terms of the Eddington luminosity (L_{Edd}),[2] is $T \propto 10^7 M^{-1/4}$ (where the temperature, T, is in K, and the mass of the BH, M, in $M_\odot$). Consequently, most of the radiation originated in the accretion disk of quasars is emitted in the UV and optical wavelengths, in contrast with the microquasar scenario, in which the radiation comes out as X-rays (see Sect. 3.1). This is actually the reason why jets in quasars, so far-distant objects, were discovered years before jets in galactic microquasars (first detection in SS 433, Margon 1984): the detection of radio jets was constrained by the development of the X-ray astronomy which would provide new stellar sources.

The main advantage of studying microquasars with respect to quasars is the time scale in which processes happen. The characteristic time of the accretion process is proportional to the mass of the BH. Whilst variations of minutes can take place in microquasars, one would have to wait thousands of years to observe the same effect in a super-massive BH (Mirabel et al. 1999). Thus, observing microquasars brings us the opportunity to study, in a possible human life-scale, the nature and origin of

---

[2]The **Eddington luminosity**, L_{Edd}, is the maximum luminosity that a body can achieve when the emitted radiation and gravitational force are balance. When the body exceeds this luminosity, the radiation pressure will overcome gravity, and material from the outer layers of the object will be forced away from it rather than falling inwards, giving rise to a very intense radiation-driven wind. This luminosity only depends on the mass of the object, $L_{Edd} = 3.2 \times 10^4 (\frac{M}{M_\odot}) L_\odot$. This luminosity is, in turn, associated with the **Eddington accretion limit**, accretion rate beyond which the L_{Edd} is overpassed.

the jets, likely related with the accretion flow. On the other side, the proximity of the galactic microquasars allow us to delve into particle acceleration inside the jets and its expected emission at very-high energies. A schematic view with the aforementioned characteristic of both microquasars and quasars is shown in Fig. 3.6.

The main components of microquasars have been introduced previously, where a detailed discussion of the companion star and the compact object can be found in Sect. 3.1. In the following, I will focus on the accretion disk and the relativistic radio-jet, correlated by the **accretion process**: the mass of the donor star affects the properties of the transferred material, like its temperature or the accretion rate, giving rise to a clear and direct effect on the accretion disk (influencing e.g. the emitted radiation from the disk or the magnetic field strength on it). Although the mechanism of production and collimation of the jets is still not perfectly understood, it is currently accepted that these streams are also powered by this accretion process, given the correlation evidences of these two mechanism (Fender et al. 2004, see Sect. 3.2.3.1). Thus, an accretion-ejection relation becomes important in these systems.

3.2.1 Accretion Disk

The movement and energy lost by friction/viscosity of the gas flow inside the accretion disk is normally described by hydrodynamic equations of the SDM, modeled by Shakura et al. (1973). This energy lost is equivalent to the binding energy at the last stable orbit, whose radio depends on the compact object. In the case of a BH, the innermost stable orbit corresponds to approximately three times the Schwarzschild radius,[3] R_{Sch}, while for the NS, it is its own surface. As mentioned before, the temperature in this orbit is defined by $T \propto 10^7 M^{-1/4}$ and hence, compact objects with tens of solar masses, as in the case of microquasars, will emit X-rays.

However, hard X-rays have been detected in several microquasars, e.g. Cygnus X-1, which will be deeply studied in the gamma-ray band in this thesis (see Chap. 4). The SDM cannot explain such high emission, unless a new component is introduced. This component is the so-called corona, hot ($T \sim 10^9$ K) plasma at the inner region of the accretion flow (Coppi 1999). Close to the compact object (at $\lesssim 100\, R_{Sch}$), the accretion disk presents low density and hence, the cooling efficiency by viscosity is low as well. This produces an increase of temperature that inflates the gas of the disk forming the corona. The hard X-ray are emitted through the Comptonization of thermal photons from the accretion disk by high-energy electrons in the corona. Hard X-rays were also speculated to be produced inside the jets in terms of Compton scattering of external photons (from the disk or the donor star).

---

[3]The **Schwarzschild radius**, a.k.a gravitational radius, is the radius of a sphere such that, assuming all its mass inside it, the escape velocity from its surface would be the speed of light. Therefore, all mass falling into the Schwarzschild radius cannot avoid the gravitational attraction of the body inside the sphere. It depends only on the mass of the object, $R_{Sch} = \frac{2GM}{c^2}$, where G is the gravitational constant and c the speed of light.

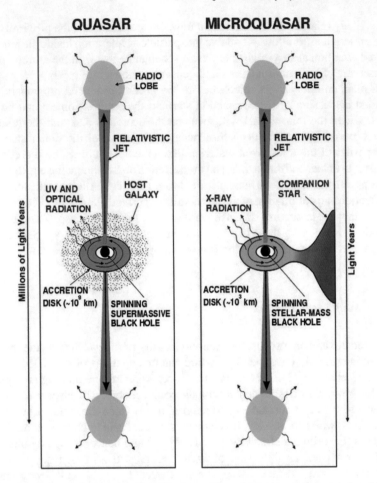

Fig. 3.6 Comparison of quasars and microquasars. Three basic components are found in both cases: a central BH, accretion disk and collimated relativistic jets. Differences are also highlighted: mass-transfer from the ISM, accretion disk of $\sim 10^9$ km emitting mostly in UV and optical and super-massive BH in the case of the quasar, and mass-transfer from a stellar companion, accretion disk of $\sim 10^3$ km emitting in X-rays and stellar-mass BH in the case of microquasars. The different distance achieved by the jets in each case is also marked (Mirabel et al. 1998)

3.2.2   Relativistic Radio-Jets

As seen before, radio-jets are highly collimated, showing opening angles of less than 15°(Mirabel et al. 1999). Normally, the angle between these outflows and the line of sight of the observer is $\gtrsim 30°$. However, this feature is purely statistical: the probability of finding jets with smaller angles is low. Microquasars whose axis of ejection form angles $\lesssim 10°$ with respect to our line of sight are known as microblazar (Fig. 3.7), again because of their scaled-up counterparts, the blazars

Fig. 3.7 Sketch of a microquasar. If the jets are aligned with the line of sight of the observer ($\lesssim 10°$), the systems are called microblazars (Mirabel 2006)

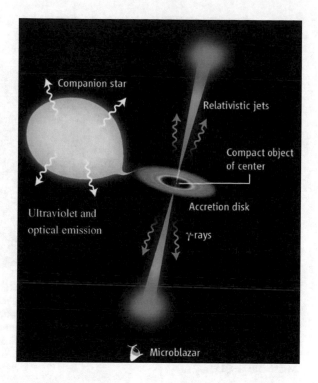

Companion star

Relativistic jets

Compact object of center

Accretion disk

Ultraviolet and optical emission

γ-rays

Microblazar

(Mirabel et al. 1999). In this kind of systems, the timescales are shortened by 2Γ, where Γ is the Lorentz factor, and the flux densities are increased by $8\Gamma^3$. Therefore, although the intensity is enhanced with respect to the microquasars, the very fast flux variability, along with the low probability to find one, make microblazars very difficult sources to detect.

Several models have been proposed along the years to describe the production and collimation of jets and the relation with the accretion process (see e.g., Blandford 1976; Blandford et al. 1977, 1982; Uchida et al. 1985, 1986; Meier 1996, among others). Currently, the most accepted one is the Magnetohydrodynamic (MHD) model, which invokes an accretor, a poloidal magnetic field and differential rotation under the assumption of high conductivity (see Fig. 3.8). A general review of the model can be found in Meier et al. (2001); here I summarize the main features that describes it:

- The plasma expelled through the jets will follow the magnetic field lines in a parallel way (along the rotation axis) without crossing them. In cases where the magnetic field is weak or the density of the plasma is high enough, the magnetic field lines will be bent back. This magnetic pressure is responsible of the outwards particle acceleration.
- Magnetic field lines repel each other, which gives rise to a perpendicular pressure. This characteristic can allow the enhancement of weak magnetic fields by bringing together several parallel lines.

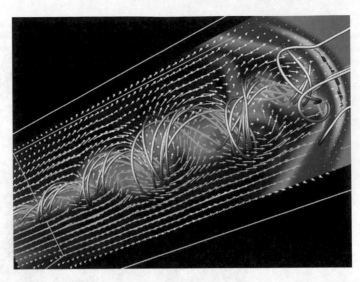

Fig. 3.8 In the axisymmetric and rotating jet (due to the conservation of the angular momentum), the plasma moves parallel to the magnetic field lines (metallic tubes). The arrows show the plasma velocity while the colored zones describes the plasma density: white indicates the highest density, blue the lowest. Plasma trapped into the magnetic field suffers Lorentz force, which can be divided into two vectors: magnetic pressure along the rotation axis, which accelerates the plasma outwards, and magnetic tension perpendicular to the former, responsible to the collimation of the streams (Meier et al. 2001)

- Magnetic field lines tend to keep straight unless other lines or forces originated in the plasma apply any effect on them. This tension is responsible for the collimation and *pinch* effect of the plasma.

The creation of the relativistic outflows in microquasars seems to be related with thick accretion disks, since no jets are detected when the system presents optically thin disks, independently of the nature of the compact object. The physical interpretation is still unclear, although one of the reasons could be the presence of a not strongly enough magnetic field, during thin accretion disk states, enable to collimate the jets. The different states and their correlation with the radio-jets are presented in the following Sect. 3.2.3.1.

Inside these outflows non-thermal emission has been detected in a broad multiwavelength range, from synchrotron radio, through IR and optical (Russell et al. 2010), up to X-rays. High-energy gamma rays from microquasars most likely originate inside the jets have been also reported. So far only two microquasars have been detected in the latter regime: the high-mass microquasar Cygnus X-3, at energies greater than 100 MeV, by Astrorivelatore Gamma ad Imagini LEggero (*AGILE*) (Tavani et al. 2009a) and *Fermi*-LAT (Fermi LAT Collaboration et al. 2009), and Cygnus X-1, at energies above 60 MeV, using *Fermi*-LAT data as reported in this thesis in Chap. 4. All these results evidence the existence of a relativistic particle population inside the streams, which are thought to be accelerated via Fermi acceleration (see Sect. 1.1.2 for a detailed description of this process). This mechanism of acceleration evokes shock waves, whose origin in microquasars can be explained by different models:

- **The jets are discrete ejections of material**: In this scenario, the shock waves are formed by the jets themselves that are ejected as blobs. These blobs can display different velocities inside the outflow and hence, interact to each other, creating internal shocks. The expected emission would be quasi-steady with a certain variability given by the blob injection (van der Laan 1966; Jamil et al. 2010).
- **The jets are continuous outflows with internal shocks**: In this case, the stream is assumed to be continuous and the shocks that produce the particle acceleration originate inside the jet by interactions of regions with different features (like densities or velocities; Kaiser et al. 2000).

3.2.3 Black Hole Microquasars

Large fraction of the microquasars in our Galaxy host a BH, which are normally highly variable X-ray sources. This variability extends to other wavelengths as well. This transient nature is related to a change in the accretion rate: they spend most of their time in low accretion periods (which leads to $L_X \lesssim 10^{33}$ erg s^{-1}), with sudden flaring periods that can last from days to months. Only a small number of this type of microquasars are persistent objects, as it is the case of Cygnus X-1 (see Chap. 4), which is always showing high accretion rate and consequently, displaying high luminosity $L_X > 10^{37}$ erg s^{-1}. This fast flux modulation complicates the study of the behavior of these systems. However, thanks to deep observations with the *RXTE* satellite, an overall view of the BH microquasars was possible, allowing to classify their X-ray states depending on flux and energy spectra changes. The course that these sources follow through all the different X-ray states is well described by the so-called *q-track* in the Hardness-Intensity Diagram (HID) given by Fender et al. (2004). In the upper plot of Fig. 3.9, we have a schematic of the HID: the X-axis corresponds to the hardness of the X-ray band, i.e. the ratio between hard and soft X-ray (the higher this value, the harder the energy spectrum would be), and the Y-axis represents the intensity of the X-ray flux. The direction of the motion is counterclockwise and give us the opportunity to define the X-ray spectra of these sources mainly by two principal states: the Hard State (HS) and the Soft State (SS).

3.2.3.1 X-Ray States

Although the use of the two canonical X-ray states on BH microquasars (the HS and the SS) is standardized, a deeper detailed X-ray emission study seems more complicated and more states have been discussed in the literature (see, e.g., Belloni et al. 2000; Gilfanov 2010). Nevertheless, these two states give us a well understanding of

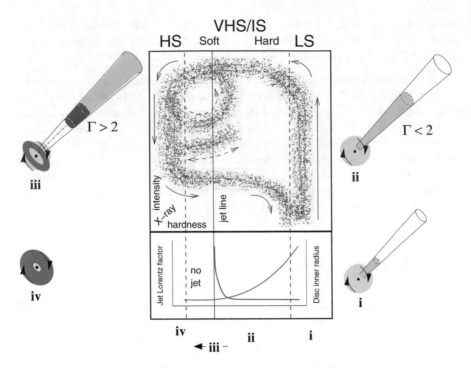

Fig. 3.9 Schematic of the jet-disk coupling model presented by Fender et al. (2004)

the systems and will be used in this thesis to discuss the behavior of the microquasars Cygnus X-3 (Chap. 5) and Cygnus X-1 (Chap. 4).

Both states are characterized by the sum of a thermal blackbody component that peaks at the keV energies and a power-law component at higher energies. The origin of the latter seems to happen in the corona, where high-energy electrons scatters (via IC) thermal photons from the accretion disk. The dominance of one or other component will define the state of the system.

- **Hard State**: It is dominated by a power-law photon distribution ($d\phi/dE \propto E^{-\Gamma}$, with $\Gamma \sim 1.4 - 1.9$) with a high-energy exponential cutoff at hundred keV, whilst the thermal component is very weak, peaking at $kT \sim 0.1$ keV. The total X-ray luminosity, L_X, can reach a few % of the L_{Edd} (Maccarone 2003). During this state, a steady radio-jet is detected, evidencing the correlation between hard X-rays and radio wavelengths. Historically, the relation between the luminosity in these two bands was $L_{radio} \propto L_X^{0.7\pm0.1}$ (Gallo et al. 2003). This value was afterwards revised by Gallo et al. (2012) differentiating two population of BH microquasars probably dependent on the accretion rate. The relation of one of them is still compatible with the former one, $L_{radio} \propto L_X^{0.63\pm0.03}$, while a new track slope is obtained, $L_{radio} \propto L_X^{0.98\pm0.08}$ (Fig. 3.10).

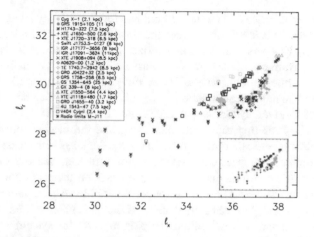

Fig. 3.10 Logarithm of radio/X-ray luminosities (l_r and l_X, respectively, in erg s^{-1}) of BH X-ray binaries by Gallo et al. (2012). This diagram evidences that the relation between radio and X-rays during the IIS follows two tendencies, with slopes 0.63 ± 0.03 and 0.98 ± 0.08. The inset shows the same information, but sources with secure distance are depicted in red and those with uncertain distance in blue

- **Soft State**: Contrary to the HS, this state is dominated by the thermal blackbody component that peaks at $kT \sim 1$ keV, emitted mainly in the inner region of the accretion disk that extends down to the last stable orbit, and a softer power-law tail (with photon index of $\Gamma \sim 2.2 - 2.7$) that extends beyond 500 keV. The X-ray luminosity in this state was never observed below 1% L_{Edd} in BH X-ray binaries. During this state, the relativistic outflows are disrupted and radio emission is undetectable.

The transition between these two states is done through the so-called Intermediate State (IS), which displays spectral properties between both of them and rapid and strong radio variability due to internal shocks inside the jets.

3.2.3.2 Disk-Jet Coupling

Instabilities in the accretion disk due to changes of the accretion rate will affect the jets radiation (and consequently, the X-ray and radio emission, as shown previously). These instabilities are accepted to be responsible of the spectral variability in BH transients. Fender et al. (2004) described these spectral changes within an unified model of coupling between disk and jet in BH binaries. The model is shown conceptually in Fig. 3.9. Besides the HID, the figure presents different sketches of the sources at each state to highlight the dominant components. The lower part of the plot shows the variation of the Lorentz factor value, Γ, and the innermost radius of the accretion disk with respect to the hardness. The different phases according to the model are:

- **Phase i**: The microquasar is in the HS but in a low-luminosity state, i.e. with low accretion rate. This state can achieve the lowest luminosities, a.k.a. *quiescence* state. It displays a persistent jet, characterized by synchrotron radio emission, and a hot corona, from where the hard X-rays arise. The last stable orbit of the accretion disk is at its furthest point from the BH. At this point, the radio and X-ray emission is related by the non-linear equation $L_{radio} \propto L_X^\alpha$, where α can be 0.63 ± 0.03 or 0.98 ± 0.08 as given by Gallo et al. (2012) (see Sect. 3.2.3.1).

- **Phase ii**: The accretion rate starts increasing and the motion through the HID becomes almost vertical, until it enters in the *hard/intermediate* state (to the left). At this moment, the spectrum starts to suffer softening because of the loss in temperature in the corona due to the IC processes on thermal photons from the disk. Thus, thermal emission is no longer negligible. At the same time, given the increase of accretion rate, the density of the gas enhances and starts to cool via synchrotron and Bremsstrahlung, helping to move the system to the SS. The corona shrinks and the innermost stable orbit approaches the BH. Due to angular momentum conservation, the closer the matter gets to the compact object, the higher the velocity is inside the jets (still $\Gamma \lesssim 2$).

- **Phase iii**: The source keeps moving to a softer state, where thermal emission from the disk becomes dominant with respect to the hard X-rays from the corona. It approaches the *jet line*, a vertical line in the HID that divides the states where the relativistic outflows are present and where they are not. In this state, the jet becomes unstable giving rise to rapid Lorentz factor increase ($\Gamma \gtrsim 2$) that originates internal shocks in the outflow before being disrupted once it enters in the SS. Between this phase and the following one, we can see a loop and a path marked with dashed lines and arrows that cross back and forward the *jet line*. This excursions, that re-activates the jets and produce flaring activity, can happen $\lesssim 10$ times.

- **Phase iv**: The microquasar finally enters the SS, where no radio-jet is detected. Therefore, the X-ray spectrum is dominated by the thermal blackbody component from the accretion disk. In this phase, the accretion rate is the highest, which extends the innermost stable orbit close to the BH. From here on, the accretion rate starts decreasing, the density of the gas in the inner part of the disk decreases as well creating the corona, and the microquasar enters the HS, where relativistic jets, in which particles can be accelerated up to gamma rays, appear again.

3.2.4 VHE Radiative Processes in Microquasars

After the discovery of relativistic jets from microquasars (Mirabel et al. 1992) and detection of non-thermal processes (from synchrotron radio up to high-energy gamma rays), VHE gamma-ray emission from this kind of systems was proposed in the literature from both leptonic (e.g. Atoyan et al. 1999; Bosch-Ramon et al. 2006) and hadronic processes (e.g. Romero et al. 2003). Bosch-Ramon et al. (2009) provide a deep review on the mechanism of particle acceleration and VHE emission of microquasars. Here I just point out the most relevant processes in the VHE regime.

3.2.4.1 Leptonic Processes

The most efficient radiative process at VHE in microquasars seems to be a leptonic one, the IC. There are different possible source photon fields according to the distance of the production site to the compact object: close to the BH, IC of thermal photons (Georganopoulos et al. 2002; Romero et al. 2002), or synchrotron photons (e.g. Bosch-Ramon et al. 2006) may be dominant. When the production region is situated inside the binary but further from the BH, the process can take place on photons from the companion star. In cases with powerful jets, VHE gamma-ray emission may also be produced in the region where the outflows interact with the environment. Other leptonic mechanisms, besides IC, have been proposed in the literature to explain lower energy emission, mainly synchrotron and relativistic Bremsstrahlung. Depending on the conditions, one mechanism would dominate over the other. Both process have been suggested to take place either at the base of the jet (see, e.g. Markoff et al. (2001) for sychrotron emission and Bosch-Ramon et al. (2006) for Bremsstrahlung) or at binary scales along the jet (e.g., Yuan et al. 2005; Bosch-Ramon et al. 2006, respectively). At the termination of the stream, synchrotron and relativistic Bremsstrahlung can also be expected as suggested by Aharonian et al. (1998) and Bordas et al. (2009).

3.2.4.2 Hadronic Processes

In the hadronic scenario, the main processes producing gamma rays are proton-proton and proton-photon interactions. The targets for the proton-proton interaction are thermal protons or ions in the jets or, more likely, in the stellar wind, whereas the photon sources are either jet synchrotron, accretion disk or stellar photospheric photons. Any of these interactions produce π^0 that, in turn, decay into two gamma rays. This hadronic collision would also produce charge pions ($\pi^\pm$) that decay into muons and neutrinos. This way, the detection of neutrinos from microquasars would be an irrefutable probe of hadronic processes taking place inside these system. Muons can, in turn, decay into electron-positron pairs and neutrinos, and these secondary pairs could be responsible for the low-energy gamma-ray emission (via synchrotron, IC or Bremsstrahlung; Orellana et al. 2007).

3.2.4.3 Radiative Processes in Low-Mass Microquasars

As we have seen, the companion star plays a key role on several models, both in leptonic and hadronic mechanisms: the stellar donor is the responsible of providing seed photons for IC or nuclei and photon targets for the proton-proton and proton-photon interactions. However, low-mass microquasars, with old and cold secondary stars, do not provide a proper environment for these processes to take place and hence, these models are only suitable in cases with high-mass companions.

Nevertheless, several models have been developed to explain gamma-ray emission from this type of systems. Zhang et al. (2015) proposed a pure leptonic model with relativistic electrons along the jet, where the dominant mechanisms ended up to be synchrotron and SSC emissions from an extended dissipation region in the jet. On the other hand, in the hadronic scenario, photopion production inside the relativistic outflow by synchrotron jet emission, given the weak wind of the secondary star, could be considered as a possible process (Levinson et al. 2001). Moreover, a simple model on proton low-mass microquasars (assuming that a significant part of the jet composition is formed by protons) was developed by Romero et al. (2008), where MeV-GeV gamma rays were predicted from these sources. TeV energies are also expected in some cases at detectable levels for the current IACTs, depending on the ratio protons/leptons. However, this simple model does not take into account rapid variability, one of the outstanding properties of these systems, as we will see with V404 Cygni in Chap. 6.

In general, gamma rays in low-mass microquasars, like V404 Cygni, are expected to be produced inside the relativistic jets or the corona by the interaction of their own matter, radiation and magnetic fields, given the lack of targets provided by the low-mass companion star (see, e.g. Bosch-Ramon et al. 2006; Vila et al. 2008; Vieyro et al. 2012).

3.2.4.4 Photon-Photon Absorption

Close to the corona, at the base of the jet, and at binary scales, the environment not only helps to produce gamma rays but extreme conditions can lead to photon-photon absorption as well. Thus, close to the compact object, thermal UV and X-ray photons coming from the accretion process produce strong absorption on GeV gamma rays via pair production. This sub-product could give rise to secondary emission from cascade, however the likely high magnetic field at the base of the jet may suppress this effect.

If VHE emission is produced at the scale of the binary system ($\lesssim R_{orb}$ from the compact object, where R_{orb} is the size of the system) in high-mass microquasars, the VHE photons will also suffer severe absorption because of the Near-infrared (NIR) stellar photon field (Orellana et al. 2007; Bednarek et al. 2007). This absorption is orbitally modulated, since it depends on the companion star-emitter-observer relative positions. This means that when the star is between the observer and the compact object (the so-called, superior conjunction of the compact object) the attenuation is at its maximum, while if the compact object is interposed between (inferior conjunction), the absorption is expected to be minimum.

3.2.5   VHE Observations of Microquasars with MAGIC

Gamma-ray emission from microquasars has been theoretically predicted for several years. However, it was not until less than a decade that gamma rays were detected from this type of systems. In 2009, two satellites *AGILE* and *Fermi*-LAT reported, for the first time, excess in energies above 100 MeV from the high-mass microquasar Cygnus X-3 (Tavani et al. 2009b; Fermi LAT Collaboration et al. 2009b). Another firmly established microquasar, composed as well with a high-mass companion, Cygnus X-1, presented a hint of steady emission reported by Malyshev et al. (2013a). Hints of transient radiation from this source were also reported by *AGILE* (Sabatini et al. 2010a; Rushton et al. 2012a; Sabatini et al. 2013a).

No VHE gamma-ray emission has been detected up to now from microquasars. Although pursued for many IACTs, this complicated task of disentangling the TeV regime in these systems is still not fulfilled, mostly due to the extremely good sensitivity required by the instruments. MAGIC has performed deep observations on microquasars since it started operation in 2004, looking to Cygnus X-1, Cygnus X-3, SS 433, GRS 1915+105 or Scorpius X-1. In 2006, MAGIC published the detection of LS I +61 303 (Albert et al. 2006), however although it was classified as microquasar at the beginning, it is currently accepted to be consistent with a pulsar wind scenario.

In this thesis, I present the latest results of two high-mass and one low-mass microquasars: Cygnus X-1, Cygnus X-3 and V404 Cygni. The former was observed in a long-term campaign from 2007 to 2014 accumulating ~100 h of good quality data. Such a deep campaign was motivated by a hint of signal detected by MAGIC in 2006 at the level of 4.1σ in the direction of this source (Albert et al. 2007). Cygnus X-3 was, at the time of starting this thesis, the best candidate for searching VHE gamma-ray emission given that it was at the moment the only microquasar detected in the gamma-ray band. It was observed following a strict follow-up observation campaign: we performed daily analysis of public *Fermi*-LAT data and according to the results on the MeV-GeV regime, MAGIC observations were triggered. V404 Cygni was observed during an extreme outburst that the system underwent on June 2015 that lasted several days. MAGIC could observed the source at its maximum activity thanks to the automatic Gamma-ray Burst procedure. With all these observations, MAGIC has been able to provide very useful information and shed light on microquasars in the VHE regime.

References

Aharonian FA et al (1998) Nature 42:579
Albert J et al (2006) Science 312:1771
Albert J et al (2007) ApJ 665:L51
Atoyan AM et al (1999) MNRAS 302:253
Bednarek W et al (2007) A&A 464:437
Belloni T et al (2000) A&A 355:271

Blandford RD (1976) MNRAS 176:465
Blandford RD et al (1977) MNRAS 179:433
Blandford RD et al (1982) MNRAS 199:883
Bondi H (1952) MNRAS 112:195
Bordas P et al (2009) A&A 497:325
Bosch-Ramon V et al (2006) A&A 447:263
Bosch-Ramon V et al (2009) Int J Mod Phys D 18:347
Camenzind M (2007) Compact objects in astrophysics: white dwarfs, neutron stars, and black holes
Chandrasekhar S (1931) ApJ 74:81
Collaboration Fermi LAT et al (2009a) Science 326:1512
Collaboration Fermi LAT et al (2009b) Science 326:1512
Coppi PS (1999) In: Poutanen J, Svensson R (eds) High energy processes in accreting black holes.
    Astronomical society of the pacific conference series, vol 161, p 375
Curtis HD (1918) Publications of lick observatory, vol 13, p 31
Fender RP et al (2004) MNRAS 355:1105
Gallo E et al (2003) MNRAS 344:60
Gallo E et al (2012) MNRAS 423:590
Georganopoulos M et al (2002) A&A 388:L25
Giacconi R et al (1962) Phys Rev Lett 9:439
Gilfanov M (2010) In: Belloni T (ed) Lecture notes in physics, vol 794. Springer, Berlin, p 17
Grimm H-J et al (2002) A&A 391:923
Gursky H (1966) S&T 32
Heger A et al (2003) ApJ 591:288
Jamil O et al (2010) MNRAS 401:394
Kaiser CR et al (2000) A&A 356:975
Levinson A et al (2001) Phys Rev Lett 87:171101
Lewin WHG et al (1995) X-ray binaries
Longair MS (2011) High energy astrophysics
Maccarone TJ (2003) A&A 409:697
Malyshev D et al (2013a) MNRAS 434:2380
Margon B (1984) ARA&A 22:507
Markoff S et al (2001) A&A 372:L25
Meier D (1996) ApJ 459:185
Meier DL et al (2001) Science 291:84
Mirabel IF et al (1992) Nature 358:215
Mirabel IF et al (1998) Nature 392:673
Mirabel IF et al (1999) ARA&A 37:409
Mirabel IF (2006) Science 312:1759
Orellana M et al (2007) A&A 476:9
Rees MJ (1998) In: Wald RM (ed) Black holes and relativistic stars, vol 79
Rees MJ (1984) ARA&A 22:471
Reig P et al (1999) MNRAS 306:100
Romero GE et al (2002) A&A 393:L61
Romero GE et al (2003) A&A 410:L1
Romero GE et al (2008) A&A 485:623
Rushton A et al (2012a) MNRAS 419:3194
Russell DM et al (2010), ArXiv e-prints
Sabatini S et al (2010a) ApJ 712:L10
Sabatini S et al (2013a) ApJ 766:83
Savonije J (1983) In: Lewin WHG, van den Heuvel EPJ (eds) Accretion-driven stellar X-ray sources,
    pp 343–366
Shakura NI et al (1973) A&A 24:337
Tavani M et al (2009a) Nature 462:620

Tavani M et al (2009b) Nature 462:620
Uchida Y et al (1985) PASJ 37:515
Uchida Y et al (1986) Can J Phys 64:507
van der Laan H (1966) Nature 211:1131
Vieyro FL et al (2012) A&A 542:A7
Vila GS et al (2008) Int J Mod Phys D 17:1903
Yuan F et al (2005) ApJ 620:905
Zhang J-F et al (2015) ApJ 806:168

Chapter 4
Cygnus X-1

4.1 History

Cygnus X-1 is one of the brightest and most studied X-ray sources in our Galaxy and a firmly established stellar-mass BH X-ray binary system. Discovered in the early stage of the X-ray astronomy (Bolton 1972), the system is located in the Cygnus region (l = 71.32° and b = +3.09°) at a distance of $1.86^{+0.12}_{-0.11}$ kpc from the Earth (Reid et al. 2011). It is comprised of a (14.81 ± 0.98) $M_\odot$ BH and a O9.7 Iab type supergiant companion star with a mass of (19.16 ± 1.90) $M_\odot$ (Orosz et al. 2011). Nevertheless, the most plausible mass range of the donor star has been recently increased to 25–35 $M_\odot$ by Ziółkowski (2014). This system is the only HMXB for which the compact object has been clearly identified as BH (Fig. 4.1).

The assumption that Cygnus X-1 ranks among the microquasars was accepted after the detection, with the VLBA instrument, of a highly collimated one-sided relativistic radio-jet that extends ∼15 mas from the source (opening angle <2c° and velocity ≥0.6c, Stirling et al. (2001). These jets are thought to create a 5 pc diameter ring-like structure observed in radio that extends up to 10^{19} cm from the BH (Gallo et al. 2005). The total power carried by these relativistic outflows is 10^{36-37} erg s^{-1} (Gallo et al. 2005; Russell et al. 2010).

The binary system moves following a slightly elliptical orbit with eccentricity of 0.018 (Orosz et al. 2011), orbital period of 5.6 days (Brocksopp et al. 1999) and an inclination angle of the orbital plane to our line of sight of 27.1 ± 0.8° (Orosz et al. 2011). The superior conjunction phase of the compact object, when the companion star is interposed between the BH and the observer (see Fig. 4.2), corresponds to phase 0, assuming the ephemerides $T_0 = 52872.788$ Heliocentric Julian Day (HJD) taken from Gies et al. (2008). As mentioned in Sect. 3.1, X-ray binaries generally suffer flux periodicity at their own orbital period. Cygnus X-1 shows this kind of modulation both in X-ray and radio wavelengths (Wen et al. 1999; Brocksopp et al. 1999; Szostek et al. 2007), which may be caused by absorption or scattering by the wind of the donnor star over the radiation emitted from the

© Springer Nature Switzerland AG 2018
A. Fernández Barral, *Extreme Particle Acceleration in Microquasar Jets and Pulsar Wind Nebulae with the MAGIC Telescopes*, Springer Theses,
https://doi.org/10.1007/978-3-319-97538-2_4

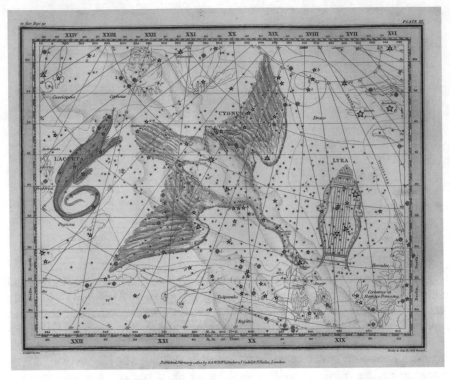

Fig. 4.1 Alexander Jamieson's Celestial Atlas representation of the Cygnus Constellation (1822). The location of Cygnus X-1 corresponds to the η symbol in the neck of the swan figure

compact object. Besides this modulation, several X-ray binary systems also present flux variations at much longer periods than their respective orbital periods. This effect is known as superorbital modulation and is thought to be caused by the precession of the accretion disk or jet (Poutanen et al. 2008). The X-ray superorbital period of Cygnus X-1 was under debate for several years. Initally, it was estimated to be $\sim$290 d by Priedhorsky et al. (1983). Later, a large number of authors claimed a superorbital periodicity of half this value, $\sim$150 d (e.g., (Brocksopp et al. 1999) or more recent Lachowicz et al. 2006). The latest results confirm again a superorbital period of $\sim$300 d, as suggested by Rico (2008) and confirmed by Zdziarski et al. (2011).

Given that is composed of a BH, Cygnus X-1 displays the two canonical X-ray spectral states of a BH transient system (see Sect. 3.2.3), the HS and the SS (Esin et al. 1998), and the course that it follows through all the different states is well defined by the HID (Fender et al. 2004). Therefore, its X-ray spectrum can be described as the sum of two components: a blackbody-like emission coming from the disk and dominant during the SS state, and a power-law tail, most likely originated due to IC of disk photons by hot thermal electrons in the corona, and dominant during the

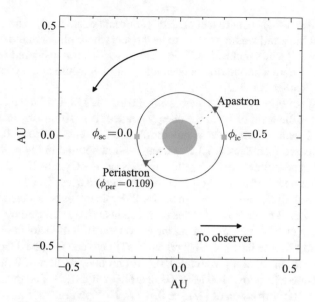

Fig. 4.2 Schematic of the Cygnus X-1 orbit where the superior ($\phi_{SC} = 0$) and inferior conjunction ($\phi_{IC} = 0.5$) are marked. The almost circular orbit (eccentricity 0.018) that follows the BH is on scale with the companion star (filled circle in the middle) of $16.4R_\odot$ (the reference included here should appear as (Orosz et al. 2011), as well as the periastron and apastron phases). Neither the inclination of the orbit with the line of sight or the longitude of the ascending node were considered here. AU stands for astronomical units. Credit: Zanin et al. (2016), reproduced with permission © ESO

HS. As shown in Sect. 3.2.3, there is a relation between radio and X-rays: whilst in the HS, microquasars display steady relativistic synchrotron jets at GHz frequencies, except for some unusual flares in Cygnus X-1 (Fender et al. 2006), during SS the radio emission is strongly quenched. However, Cygnus X-1 is a persistent X-ray source never fully disk-dominated, i.e. even during its SS the system presents a strong power-law component and evidences of an unresolved compact jet during this state (Rushton et al. 2012). Nevertheless, this jet is 3–5 times weaker than the one observed during the HS that reached $0.6c$ (Gallo et al. 2005). Thus, there is a constant level of radio emission around 10–15 mJy, that extends with no cutoff, up to the IR band, where the contribution from the O9.7 Iab type donor dominates, hindering the measurement of such cutoff.

Observations with COMPTEL when Cygnus X-1 remained in the SS suggested, for the first time, the existence of non-thermal component beyond MeV (McConnell et al. 2002). This result gave rise to an increase of the interest for this source in the gamma-ray regime. Nevertheless, observations with *INTEGRAL* excluded the existence of this MeV tail in the SS, but probed, in turn, the presence of such non-thermal hard emission during the HS, when the jets were present (Rodriguez et al. 2015). *INTEGRAL*-IBIS also reported a hard tail in the HS which was shown to be polarized in the energy range of 0.4–2 MeV at a level of ~70% with a polarization angle of $(40.0 \pm 14.3)°$ (Laurent et al. 2011; Jourdain et al. 2012). The origin site of this polarized MeV tail was suggested to be the jets where ultra-relativistic electrons would produce it via synchrotron. The corona was also considered as source of this radiation, where a population of secondary leptons would emit synchrotron soft gamma rays (Romero et al. 2014).

Steady high-energy gamma-ray emission during the HS was hinted by Malyshev et al. (2013b) at the level of 4σ in the energy range of 0.1–10 GeV by using 3.8 years of *Fermi*-LAT data. At the time of this thesis, the 7.5 years of Pass 8 *Fermi*-LAT data were released (see Sect. 2.6.2.1). Given the hint spotted in the past, with more available data and better sensitivity, the analysis of this Cygnus X-1 data set could clearly provide new information on accreting X-ray binaries. The data used and the results of such analysis can be found in Sect. 4.2. Besides this persistent emission, the source underwent 3 preceding episodes of transient activity detected by *AGILE*. The first two flaring events occurred during the HS on the 16th of October 2009, with an integral flux of $(2.32 \pm 0.66) \times 10^{-6}$ ph cm^{-2} s^{-1} between 0.1 and 3 GeV (Sabatini et al. 2010), and on the 24th March 2010, with an integral flux of 2.50×10^{-6} ph cm^{-2} s^{-1} for energies above 100 MeV (Bulgarelli et al. 2010). The third one, on the 30th of June 2010 with a flux of $(1.45 \pm 0.78) \times 10^{-6}$ ph cm^{-2} s^{-1} also for energies above 100 MeV (Sabatini et al. 2013), took place during the IS when the source was leaving the HS but just before an atypical radio flare (Rushton et al. 2012). All these episodes lasted only 1–2 days. An independent analysis performed by Bodaghee et al. (2013) using 3.6 years of *Fermi*-LAT data confirmed, at the level of 3–4σ, transient emission from Cygnus X-1, although not coincident with the *AGILE* flares (between one and two days before the event reported by Sabatini et al. 2010).

MAGIC observed the source in the past for a total of 40 h, spanning 26 nights between June and November of 2006. During that period, the observations were carried out with the stand-alone MAGIC telescope, MAGIC I. Although no significant excess for steady gamma-ray emission using the all data sample was found, during the daily basis analysis a hint on the 24th of September 2006 (MJD = 54002.96), corresponding to an orbital phase of 0.9 (i.e. close to the superior conjunction of the compact object) was spotted (Albert et al. 2007). This search yielded an evidence of gamma rays at 4.9σ (4.1σ after trials) in an effective time of 79 min. This excess took place at the maximum superorbital modulation of the source and simultaneously with the rising edge of a hard X-ray flare detected by *INTEGRAL*, *Swift*-BAT and *RXTE*-ASM (Malzac et al. 2008). The energy spectrum computed for this day is well defined by a simple power law of $d\phi/dE = (2.3 \pm 0.6) \times 10^{-12} (E/1\text{TeV})^{-3.2\pm0.6}$ TeV^{-1} cm^{-2} s^{-1}. The VERITAS Collaboration also observed Cygnus X-1 on 2007 without any significant detection (Guenette et al. 2009) and therefore, former MAGIC results were the first experimental hint of VHE emission from a stellar BH binary. Consequently, both HE and VHE hints triggered a deep campaign on Cygnus X-1, whose results are shown in this chapter.

4.2 Fermi-LAT Analysis

4.2.1 Data Selection

For this analysis, we used $\sim$7.5 years of **Pass 8** data, from the 4th of August 2008 (MJD 54682) to the 2nd of February 2016 (MJD 57420). As shown in Sect. 2.6.2.1, the use of **Pass 8** data allowed us to increase the energy range and cover from 60 MeV up to 500 GeV. These results were obtained through two independent methods that worked as cross-check for the results: on one hand, I analyzed the sample following the standard FERMI SCIENCE TOOLS,[1] and on the other, results were obtained using the *Fermipy*[2] package, a set of python tools recently released to automatize the **Pass 8** analysis of *Fermi*-LAT data. In both cases, and in order to properly compare results, the cuts applied to the data were the same. It was selected "P8R2_SOURCE" class photons from the LAT archive, with the aim of looking in the widest possible energy range (60 MeV–500 GeV), that were located within a 30° acceptance cone from Cygnus X-1, assuming the coordinates $RA_{J2000} = 19h{:}58m{:}21.676s$, $Dec = +35°{:}12m{:}5.78s$ (van Leeuwen 2007). Only "SOURCE" class events were used, which correspond to those with high probability of being photons. Therefore, the corresponding "P8R2_SOURCE_v16" IRFs were taken, which define the response of the satellite to gamma rays with a certain energy (defined by the EDISP event-type) and arrival direction (whose quality is divided in PSF event-type) in the coordinates of the instruments for the given period. When performing the analysis with the standard FERMI SCIENCE TOOLS, I used those events with PSF event-type FRONT+BACK and cross-checked results by selecting the events separately from the FRONT and BACK track. However, with **Pass 8** the photon class events can be subdivided into quartiles regarding the quality of their PSF and energy reconstruction. Thus, the analysis with the *Fermipy* tools was computed for the four PSF event-type separately and combining the output results with a joint likelihood fit. Given the possible contamination by the albedo gamma rays from the Earth, those photons whose reconstructed direction angle with respect to the instrument was larger than 90°, 85°, 75° and 70° for each of the four PSF quartiles (the tighter, the better the PSF is) were not included in the study. In the case of the standard analysis, I fixed the conservative value of 90°. On the other side, no cut was applied to the rocking angle, since the analysis is not very sensitive to the Earth Limb and data in which the *Fermi*-LAT satellite was in pointing mode to the Galactic-Center was also included. By analyzing the PSF event-type separately with the *Fermipy* tools, the best possible angular resolution was achieved, e.g. $\sim$0.5° at 1 GeV. Both analysis, using standard FERMI SCIENCE TOOLS and *Fermipy* tools were probed to be compatible.

---

[1] http://fermi.gsfc.nasa.gov/ssc/data/analysis/software/.
[2] http://fermipy.readthedocs.org/en/latest/.

4.2.2   Model

To create a model that defines the emission coming from the region around Cygnus X-1, we used $14° \times 14°$ RoI in galactic coordinates. The model needs to account for the galactic diffuse emission as well as the isotropic contribution, mainly composed by extragalactic diffuse radiation and cosmic rays. The corresponding background models, utilized in this analysis, were $gll\_iem\_v06.fits$, for the galactic contribution, and $iso\_P8R2\_SOURCE\_V6\_PSFx\_v06.txt$ (where $x = 0, 1, 2$ and 3), for the isotropic one. As point-like background sources, we first selected all those contained within $22°$ radius from the third LAT catalogue (3FGL, Acero et al. 2015), which used ~ 4 years of Pass 7 Reprocessed LAT data. All spectral parameters from sources further away more than $14°$ from the Cygnus X-1 location were fixed, given the small contribution to the total amount of photons arriving from the target nominal position. We left free the flux normalization of sources between $7°$ and $14°$ and that of extremely bright sources (with significance above 100σ), as well as the flux normalization of the diffuse components (both galactic and isotropic). The closest sources, those placed at less than $7°$ from Cygnus X-1, preserved all their spectral parameters free to vary in the maximum likelihood fit performed with the *gtlike* function.

The significance is determined through the TS value, defined as $TS = -2\ln (\mathcal{L}_0/\mathcal{L}_1)$, where $\mathcal{L}_0$ is the likelihood value for the model without including Cygnus X-1 (the so-called null hypothesis) and L_1 is the likelihood including it. This way, the TS will be maximized when the likelihood of the model with our source is maximized as well. For all the TS maps we obtained around the microquasar, we adopted a power-law function to describe the Cygnus X-1 spectrum with a photon index of 2.5, while all the point-like background sources remained fixed.

The best localization of Cygnus X-1 was searched for energies above 1 GeV (assuring thus a good angular resolution), using *Fermipy*. The method consists in two steps: from a $4° \times 4°$ TS map centered around Cygnus X-1, the algorithm looks for the location with the maximum TS value, and afterwards it performs a full likelihood fit around this peak of TS to provide the best position. During this process, the flux normalization of extremely bright source are free to vary.[3]

The first TS residual maps obtained were not flat, i.e. the background emission was not well described including only the 3FGL sources. While the diffuse and isotropic contribution is not expected to change between the period in which the 3FGL catalogue was released and the time of our analysis, the larger (~ 7.5 years) Pass 8 data sample allowed to improve the sensitivity and hence new point-like sources arised. We found in this first model 7 new hotspots with TS above 25 (for the full energy range 0.6–500 GeV), a part from an already clear excess at the position of Cygnus X-1 (see Fig. 4.3).

---

[3]http://fermipy.readthedocs.io/en/latest/advanced/localization.html.

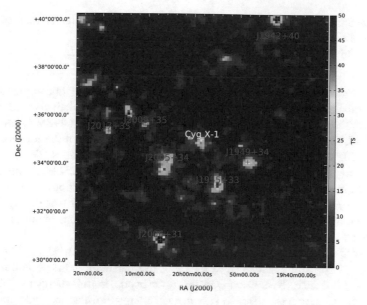

Fig. 4.3 $5° × 5°$ TS map centered at Cygnus X-1 for energies above 1 GeV in which the point-like background model is only defined by sources from the 3FGL catalogue. Credit: Zanin et al. (2016), reproduced with permission © ESO

In order to account for the contribution of these candidates, we included them in the model by assuming, in most of the cases, that their spectra follow a power-law function. General features of these sources are:

- **J1942+40**: This source presented a TS = 55 between 0.6–500 GeV. The excess corresponds to the location $RA_{J2000} = 19h:42m:7s$, $Dec = +40°:14m:7s$, most probably associated to the open cluster NGC 6819, where Gosnell et al. (2012) detected several X-ray sources with the *XMM-Newton* Observatory.
- **J1949+34**: This source displayed the lowest significance with a TS = 35 also in the full energy range. The coordinates associated to it are $RA_{J2000} = 19h:49m:7s$, $Dec = +34°:15m:44s$.
- **J1955+33**: It showed strong a excess with TS = 90 (between 0.6–500 GeV) at $RA_{J2000} = 19h:55m:10s$, $Dec = +33°:18m:34.8s$.
- **J2005+34**: The LAT excess (centered at $RA_{J2000} = 19h:42m:7s$, $Dec = +40°:14m:7s$), came to a TS = 49 in the full energy range.
- **J2006+31**: This is an extremely bright new LAT source with TS = 115 at $RA_{J2000} = 20h:06m:12.8s$, $Dec = +31°:02m:38.3s$. It is spatially coincident with the 164 ms radio pulsar PSR J2006+3102 (Nice et al. 2013). Just for this case, the spectrum was better described (at more than 3σ level) by a LogParabola instead of power-law function.

- **J2009+35**: This source presented a TS = 48 for energies greater than 60 MeV centered at RA_{J2000} = 20h:09m:57.8s, Dec = +35°:44m:48.6s.
- **J2017+35**: Its excess reached TS = 90 at RA_{J2000} = 20h:17m:25s, Dec = +35°: 26m:5s.

The coordinates given for these 7 new sources were estimated for energies above 1 GeV to consider events with better angular resolution, with a statistical uncertainty of ~0.2°. On the other hand, we found that the source 3FGL J2014.4+3606, associated to the SNR G73.9+0.9, presented a mismatch of 0.24° between the location given in the 3FGL catalogue and the centroid obtained in our TS maps. The new coordinates were set to RA_{J2000} = 20h:13m:33.8s, Dec = +36°:11m:54.0s. In the model, the spectrum of this source is described by a power-law function, since the more complex LogParabola function, suggested by Zdziarski et al. (2016b) using Pass 8 data, was not favoured. Taken into account all these modifications in our model, the background was then well defined.

The SED was computed for the entire energy range, i.e. from 60 MeV up to 500 GeV, using 7 logarithmically spaced bins. For this calculation, the photon index of Cygnus X-1 is fixed to the one obtained during the overall fit (full energy range), while the flux normalization is let free to vary in each energy bin. The spectral parameters of the rest of the point-like background sources are fixed by the overall fit too. ULs (at the 95% C.L.) are calculated if the TS in one bin does not reach at least 4, i.e. if the significance is less than 2σ. The SEDs were obtained with the *Fermipy* software package.

In order to compute a lightcurve of the more than 7 years of data, which is calculated by applying the maximum likelihood fit in a daily basis, we used the FERMI SCIENCE TOOLS for energies above 100 MeV. With this energy range, the results reported here can be compared with the previous flaring activities reported from Cygnus X-1 (Sabatini et al. 2010, 2013; Bodaghee et al. 2013). In this case, ULs at the 95% C.L. are computed when the TS in 1 day-bin was lower than 9 (significance $<3\sigma$). If the TS was higher, an integral flux for energies between 100 MeV and 20 GeV was calculated. The former maximum value of 20 GeV was constrained by the results from the SED, as shown in the next subsection.

4.2.3   Results

The model (including the 3FGL and the new hotspots as point-like background emitters) leads us to claim the first firmly gamma-ray detection at the location of Cygnus X-1 with a TS = 53, from 60 MeV to 500 GeV. The microquasar is also detected at energies above 1 GeV at the level of TS = 31. With the previously mentioned *Fermipy* algorithm to calculate the position of the source, the best coordinates for Cygnus X-1 were set to RA_{J2000} = 19h:58m:56.8s, Dec = +35°:11m:4.4s, which shows an offset of 0.05° with respect to the van Leeuwen (2007) location, but still compatible with Cygnus X-1 within the statistical uncertainties of 0.2°.

Table 4.1 Intervals in MJD of the HS and the SS periods of Cygnus X-1. Credit: Zanin et al. (2016), reproduced with permission © ESO

HS	SS
54682–55375	55391–55672
55672–55790	55797–55889
55889–55945	55945–56020
56020–56086	56086–56330
56718–56753	56338–56718
56759–56839	56839–57009
56848–56852	57053–57103
57009–57053	57265–57325
57103–57265	
57325–57420	

Given the former hinted dependency on the X-ray spectral state, we also performed the analysis of the source dividing the sample according to its HS and SS periods. These periods were defined with the public available *Swift*-BAT (15–50 keV) data, following the criterion given by Grinberg et al. (2013): above 0.09 counts cm^{-2} Cygnus X-1 stays in the HS+IS and below in the SS. It is worth mentioning that, taken into account the short duration (around days) of the transition periods between these two main states, the inclusion of the IS into the HS period cannot alter the results. The division made into these two main periods is shown in Table 4.1. The total amount of time in each subsample is very similar, 3.6 years for the HS and 3.7 years for the SS.

This analysis results on a clear detection of Cygnus X-1 during the HS with a TS = 49 for the full energy range, which leads to an energy flux of $(7.7 \pm 1.3) \times 10^{-6}$ MeV cm^{-2} s^{-1}. However, no significant gamma-ray emission from the source is detected during its spectrally soft state, which showed a very low TS = 7. For the latter, integral ULs for energies greater than 60 MeV and a 95% C.L. was set at 5.4×10^{-6} MeV cm^{-2} s^{-1}. This result evidences the correlation between the GeV emission and the hard X-rays, previously hinted by Malyshev et al. (2013a) and confirmed, just after the time of the publication of these results, by an independent analysis done by Zdziarski et al. (2016a). In Fig. 4.4, the TS maps for energies above 1 GeV for each of the X-ray states are shown.

Due to the variability shown by Cygnus X-1, like the orbital modulation in X-rays or radio wavelengths, the HS data sample was also analyzed according to the orbital phase. However, due to the low statistics, the division was performed in only two bins, aiming to center the data on the superior conjunction of the compact object (at phases between $\phi > 0.75$ and $\phi < 0.25$) and the inferior conjunction ($\phi > 0.25$ and $\phi < 0.75$), which are critical phases given the almost circular orbit of the system. The source is detected around the superior conjunction at TS = 31, while it remains undetectable (TS = 10) during the inferior conjunction (Fig. 4.5). Even though the statistics are low to study deeper this excess, the strong difference between the two bins (with the same effective time) can be understood as a hint of gamma-ray orbital

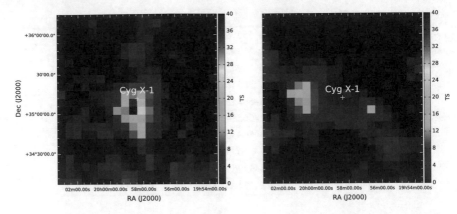

Fig. 4.4 $2° \times 2°$ TS maps for energies above 1 GeV centered at the position of Cygnus X-1 (white cross). *Left panel*: TS map corresponding to the subsample data during the HS, where the detection (TS=49) is evident. *Right panel*: TS map corresponding to the subsample data during the SS. Credit: Zanin et al. (2016), reproduced with permission © ESO

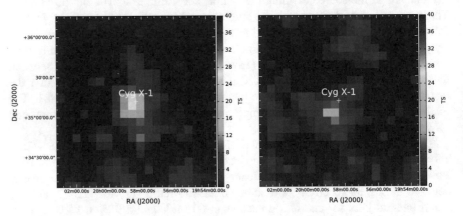

Fig. 4.5 Phase-folded $2° \times 2°$ TS maps for energies above 1 GeV centered at the position of Cygnus X-1 (white cross) when the source was in the HS. *Left panel*: TS map for phases around the superior conjunction of the compact object ($\phi > 0.75$ and $\phi < 0.25$). *Right panel*: TS map for phases around the superior conjunction of the compact object ($\phi > 0.25$ and $\phi < 0.75$). Credit: Zanin et al. (2016), reproduced with permission © ESO

modulation. The energy flux during the superior conjunction is $(7.6 \pm 1.7) \times 10^{-6}$ MeV cm^{-2} s^{-1}.

The SED obtained for the HS is well defined by a simple power-law function with a photon index $\Gamma = (2.3 \pm 0.1)$ and a flux normalization of $f_0 = (5.8 \pm 0.9) \times 10^{-13}$ MeV^{-1} cm^{-2} s^{-1} at a pivot energy of 1.3 GeV (see Fig. 4.6). It extends from 60 MeV up to $\sim$20 GeV, energy from which ULs were set. A broken power-law function was not favoured against a simple power-law: the improvement was not statistically significant, ΔTS < 2. The spectrum computed for the phase-folded analysis yielded

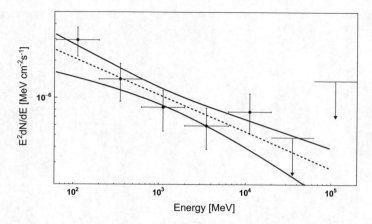

Fig. 4.6 SED of Cygnus X-1 during the HS periods. It is well described by a power-law function with photon index $\Gamma = (2.3 \pm 0.1)$ and extends up to ~ 20 GeV. Credit: Zanin et al. (2016), reproduced with permission © ESO

photon indices compatible within 1σ with the $\Gamma = 2.3$, in both superior and inferior conjunctions. The flux normalization for the superior conjunction was $f_0 = (5.7 \pm 1.3) \times 10^{-13}$ MeV^{-1} cm^{-2} s^{-1}.

Given the rapid variation of the flux level previously reported by MAGIC on a timescale of hours (Albert et al. 2007), as well as the flaring activity hinted by Sabatini et al. (2010, 2013), Bodaghee et al. (2013), we carried out daily basis analysis of the ~ 7.5 years of data in an energy range between 100 MeV and 20 GeV (Fig. 4.7). This search yielded no gamma-ray significant excess in any of the nights. All days with TS > 9 are listed in Table 4.2 and among them, only MJD 55292 is coincident with one transient event formerly reported by Bodaghee et al. (2013). No hint was either highlighted any day around the flaring events reported by *AGILE* (Sabatini et al. 2010, 2013; Bulgarelli et al. 2010). This apparent discrepancy can be explained based on the different exposure time and off-axis angle distance both satellites, *Fermi*-LAT and *AGILE*, presented during Cygnus X-1 observations, as discussed by Munar-Adrover et al. (2016) for the case of AGL J2241+4454.

4.2.4 Discussion

As it was discussed in Sect. 3.2.4.3, several mechanisms have been used to explain HE and VHE gamma-ray emission from microquasars. The IC scattering on seed photons would be dominant if the HE radiation does not come from the lower part of the jet. Along the outflow, different photon fields can work as target for this process. At a distance of $\sim 10^{11}$ cm from the compact object, the dominant photon field is the stellar radiation one (Romero et al. 2014), defined as $\omega_\star = L_\star / 4\pi (R_{orb}^2 + Z^2)c$,

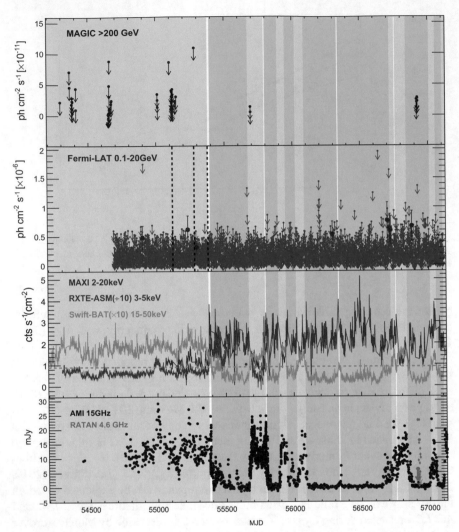

Fig. 4.7 *From top to bottom*: daily MAGIC integral ULs for $E > 200$ GeV assuming a power-law function with photon index $\Gamma = 3.2$, HE gamma rays from *Fermi*-LAT, hard X-ray (*Swift*-BAT, $\times 10$ counts s^{-1} cm^{-2} in the 15–50 keV range), intermediate-soft X-ray (*MAXI*, in counts s^{-1} in the 2–20 keV range), soft X-ray (*RXTE*-ASM, counts s^{-1} divided by 10 in the 3–5 keV range), and finally, radio integral fluxes from AMI at 15 GHz and RATAN-600 at 4.6 GHz. In the HE pad, daily fluxes with $TS > 9$ are displayed as filled black points while days with $TS < 9$ are given as 95% C.L. ULs. Dashed lines, in the same pad, correspond to *AGILE* alerts. For convenience, an horizontal green dashed line in *Swift*-BAT plot is displayed at the limit of 0.09 counts cm^{-2} s^{-1}, above which the source can be considered to be in the HS and below which it is in the SS (Grinberg et al. 2013). This distinction between X-ray states is also highlighted by the color bands: gray bands correspond to the HS+IS and blue ones to the SS periods. White bands correspond to transitions between these two main X-ray spectral states which cannot be included within the HS periods. Credit: Ahnen et al. (2017)

Table 4.2 Days, also in MJD, with a TS > 9 (significance ≥3σ) obtained during the daily basis analysis in the energy range between 100 MeV and 20 GeV. The fourth and fifth columns correspond to the *Fermi*-LAT flux and the X-ray state of the source in each day, respectively. Credit: Zanin et al. (2016), reproduced with permission © ESO

Date		TS	*Fermi*-LAT flux (10^{-7} photons cm^{-2} s^{-1})	X-ray state
(yyyy mm dd)	(MJD)			
2009-03-05	54895	10.3	4.8 ± 2.0	HS
2010-02-02	55229	10.5	6.2 ± 2.3	HS
2010-04-06	55292	12.2	3.1 ± 1.1	HS
2012-12-31	56292	9.2	5.3 ± 2.2	SS
2014-02-10	56698	9.7	7.7 ± 3.1	SS
2014-03-01	56717	10.5	6.3 ± 2.5	SS
2014-03-06	56722	9.4	6.0 ± 2.5	HS
2014-08-08	56877	10.2	6.7 ± 2.5	SS
2015-05-26	57168	9.6	4.5 ± 1.8	HS

where $L_\star = 7 \times 10^{39}$ erg s^{-1} is the luminosity of the supergiant star in the Cygnus X-1 system (Orosz et al. 2011), $R_{orb} = 3 \times 10^{12}$ cm is the orbital distance and Z is the distance from the compact object along the jet. In comparison, the soft X-rays (1–20 keV) produced in the accretion disk due to viscosity forces, with a density of $\omega_{softXrays} = L_{softXrays}/4\pi Z^2 c$, present a luminosity, $L_{softXrays}$, between 10^{36} and 2×10^{37} erg s^{-1} (depending on the model, Di Salvo et al. 2001). Further away from the BH, the photon density from the companion star diminishes, but gamma rays can be produced in the shocks between the relativist outflow and the surrounding medium. This kind of interaction is believed to take place in Cygnus X-1 due to the inflate ring-like structure detected in radio by Gallo et al. (2005). This structure extends up to 10^{19} cm from the BH and hence, this distance is assumed to be the maximum extension of the jets in Cygnus X-1.

However, inside the binary, the gamma rays will suffer severe absorption due to photon-photon collision. In the case of GeV photons, this absorption is heavier close to the base of the jet where the UV or soft X-ray photon density coming from the disk is high. For TeV photons (in the range of the MAGIC telescopes observations), this effect is greater at distances in which the stellar radiation field is dominant, given the high contribution of IR emission.

Taken this into account and the fact that we do detect GeV gamma rays from Cygnus X-1, we can calculate the minimum distance from the BH to avoid pair production between the GeV photons and soft X-rays ($\sim$1 keV) from the disk, following Akharonian et al. (1985) approach:

$$R \geq 6 \times 10^6 \left(\frac{d}{1\text{kpc}}\right)^2 \left(\frac{F_{X-rays}}{10^{-2} \quad \text{keV}^{-1}\text{cm}^{-2}\text{s}^{-1}}\right) \tag{4.1}$$

where $d = 1.86$ kpc is the distance to Cygnus X-1 (Reid et al. 2011) and $F_{X-rays} = 1.6 \times 10^{-9}$ erg cm^{-2} s^{-1} is the de-absorbed flux at 1 keV given by Di Salvo et al. (2001). Considering these values, the production site of GeV gamma rays must be larger than R $> 2 \times 10^9$ cm from the BH to avoid being absorbed. Given that the radius of the corona is around 20–50 R$_g$, where R$_g$ is the gravitational radius, i.e. $\sim 5 - 10 \times 10^7$ cm, we can exclude the corona as the emission region for the observed gamma rays, which disfavors advection-dominated accretion flow (ADAF) models. Moreover, the GeV emission should be produced, not only outside the corona, but most likely inside the jets. This conclusion is based on the exclusive detection of the system during HS, when the relativistic jets are present.

If the hint of orbital modulation reported here is finally confirmed, we could exclude the jet-medium interaction regions, given that it is not affected by orbital variability and hence, the emission should originate from the jets themselves. Thus, we can establish an UL on the maximum distance with respect to the compact object at $Z < 10^{13}$ cm. Additionally, this type of modulation is only expected in case the GeV emission arises from anisotropic IC scattering on stellar radiation (Jackson 1972; Aharonian et al. 1981; Zdziarski et al. 2013; Khangulyan et al. 2014). IC or SSC mechanisms on thermal photons from the accretion disk can be discarded since no orbital modulation is expected (assuming no additional sources of variability). As mentioned before, the stellar photon field becomes dominant at $Z > 10^{11}$ cm from the BH, which allows us to constrain the GeV gamma-ray emitter location to 10^{11}–10^{13} cm. This conclusion is compatible with the hydrodynamic simulations of the interaction between the stellar wind and Cygnus X-1-like jets with power $\sim 10^{36-37}$ erg s^{-1} (Perucho et al. 2008; Yoon et al. 2016).

To achieve gamma rays between 60 MeV and 20 GeV, the energy of the primary electrons needs to be at least of the order of several tens of GeV, which means that the IC scattering takes place in the Thomson regime (where $E_\gamma E_{e^-} << m_e^2 c^4$, as explained in Sect. 1.2.1.5). In order to accelerate the electrons to those energies within the stellar photon field, a moderate magnetic field would be enough, as suggested by Khangulyan et al. (2008), i.e. B $\sim 10^{-2}$G $\times \eta$, where η is the acceleration efficiency. If we now assume that the same population of electrons that produces the GeV gamma-ray emission is responsible of the synchrotron radiation detected in the jets at lower energies, we can constrain the magnetic field strength by the ratio of the X-ray and HE gamma-ray luminosities observed:

$$\frac{B^2}{8\pi} = \omega_\star \frac{L_{X-ray}}{L_{GeV}} \tag{4.2}$$

The energy flux obtained in this work during the HS for the full energy range, $(7.7 \pm 1.3) \times 10^{-6}$ MeV cm^{-2} s^{-1}, corresponds to a luminosity of $L_{GeV} = 5 \times 10^{33}$ erg s^{-1}, assuming the distance of 1.86 kpc to the system. Note that this value is a few orders of magnitude smaller than the total power emitted by the jets, 10^{36-37} erg

s^{-1} (Gallo et al. 2005; Russell et al. 2010). In this case, we considered the $L_{X-ray} = 2.2 \times 10^{37}$ erg s^{-1} between 20 and 100 keV. Thus, for a distance of $Z = 10^{12}$ cm, we obtain a maximum magnetic field strength of $\sim$2 kG, which decreases down to 700 G at the edge of our production region ($Z = 10^{13}$ cm).

4.3 MAGIC Analysis

4.3.1 Observations and Data Analysis

MAGIC observed Cygnus X-1 from July 2007 to October 2014. Therefore, observations were carried out in stand-alone mode with MAGIC I (Aliu et al. 2009), with pre-upgrade telescopes (Aleksić et al. 2012b) and with post-upgrade stereo system (Aleksić et al. 2016a, b). Further information on the performance of each period can be found in Sect. 2.4.

The analysis was performed using the MARS software described in Sect. 2.4.3. Integral and differential flux ULs were computed making use of the full likelihood analysis developed by Aleksić et al. (2012a), which takes into account the different IRFs of the telescopes along the years, assuming a 30% systematic uncertainty.

At La Palma, Cygnus X-1 culminates at a zenith angle of 6°. Observations, performed up to 50°, were carried out in stand-alone mode (with just MAGIC I) from July 2007 to summer 2009, and in stereoscopic mode from October 2009 up to October 2014. Two data taking modes were used: the false-source tracking mode, *wobble-mode*, and the *ON/OFF mode* (for deeper information see Sect. 2.4.2.1). In the former one, for this analysis, MAGIC pointed at two and four different positions situated 0.4° away from the source to evaluate the background simultaneously (Fomin et al. 1994). In the latter mode, the background sample, observed separately from the *ON region*, was recorded under same conditions (same epoch, zenith angle and atmospheric conditions) as for the *ON data* but with no candidate source in the FoV. The total Cygnus X-1 data sample recorded by MAGIC amounts to $\sim$97 h after data quality cuts distributed over 53 nights between July 2007 and October 2014. As pointed out before, the whole data sample extends over five yearly campaigns, characterized by different performances of the telescopes. Because of this, each epoch was analyzed separately with appropriate MC-simulated gamma-ray events. The details of the observations for each campaign are summarized in Table 4.3. For convenience, the following code is used in the table to describe the different observational features: STEREO stands for stereoscopic mode while MONO is used when only MAGIC I was operating. In the latter, the subscript specifies the observational mode: *ON/OFF* or *wobble mode*. In STEREO, only *wobble mode* was used, so the subscript is used to specify whether the observations were taken before (pre) or after (post) the MAGIC upgrade.

Different criteria to trigger observations were used during the campaign to optimize observations, aimed at observing the system in a given state, the HS, similar

Table 4.3 *From left to right*: date of the beginning of the observations in calendar and in MJD, effective time after quality cuts, zenith angle range, X-ray spectral state and observational conditions. Horizontal lines separate different observational modes along the campaign. During MJD 54656, 54657 and 54658, data under different observational modes were taken. Credit: Ahnen et al. (2017)

Date		Eff. time	Zd	Spectral	Obs.
[yyyy mm dd]	[MJD]	[h]	[deg]	State	conditions
2007 07 13	54294	1.78	6.5–17.0	HS	MONO$_{wobble}$
2007 09 19	54362	0.71	25.1–50.8		
2007 09 20	54363	1.43	21.3–40.9		
2007 10 05	54378	0.85	6.5–26.4		
2007 10 06	54379	1.85	6.4–25.8		
2007 10 08	54381	1.95	17.8–43.1		
2007 10 09	54382	0.77	9.6–34.3		
2007 10 10	54383	2.26	6.9–33.3		
2007 10 11	54384	0.76	11.1–33.3		
2007 11 05	54409	0.58	34.2–48.6		
2007 11 06	54410	0.96	20.0–33.2		
2008 07 02	54649	4.24	6.5–30.1	HS	MONO$_{on/off}$
2008 07 03	54650	3.26	6.5–30.3		
2008 07 04	54651	4.27	6.5–30.1		
2008 07 05	54652	4.15	6.4–36.1		
2008 07 06	54653	3.75	6.5.36.3		
2008 07 07	54654	3.69	6.5–37.4		
2008 07 08	54655	3.94	6.5–34.1		
2008 07 09	54656	3.06	6.5–33.8		
2008 07 10	54657	2.89	6.5–36.8		
2008 07 11	54658	1.18	6.5–30.1		
2008 07 09	54656	0.33	28.5–33.5	HS	MONO$_{wobble}$
2008 07 10	54657	0.39	21.5–36.5		
2008 07 11	54658	0.32	14.8–19.6		
2008 07 12	54659	2.51	6.5–31.0		
2008 07 24	54671	0.62	13.0–19.6		
2008 07 25	54672	0.63	8.4–14.4		
2008 07 26	54673	0.84	6.5–9.1		
2008 07 27	54674	0.30	9.5–12.7		
2009 06 30	55012	3.50	6.0–30.0		
2009 07 01	55013	2.63	6.0–30.0		
2009 07 02	55014	1.83	6.0–30.0		
2009 07 05	55017	0.22	25.0–35.0		

(continued)

Table 4.3 (continued)

Date		Eff. time	Zd	Spectral	Obs.
[yyyy mm dd]	[MJD]	[h]	[deg]	State	conditions
2009 10 08	55112	0.26	6.1–14.3	HS	STEREO$_{pre}$
2009 10 10	55114	0.67	20.0–32.6		
2009 10 11	55115	2.03	6.0–40.4		
2009 10 12	55116	2.34	6.9–42.4		
2009 10 13	55117	0.95	26.0–41.2		
2009 10 14	55118	1.98	7.5–40.0		
2009 10 16	55120	1.37	7.5–40.0		
2009 10 17	55121	0.96	7.5–40.0		
2009 10 18	55122	1.60	7.5–40.0		
2009 10 19	55123	0.68	7.5–40.0		
2009 10 21	55125	1.99	7.5–40.0		
2009 11 06	55141	0.37	7.5–40.0		
2009 11 07	55142	0.64	7.5–40.0		
2009 11 13	55148	0.89	7.5–40.0		
2010 03 26	55281	0.78	38.5–50.0		
2011 05 12	55693	1.35	12.3–42.1		
2011 05 13	55694	1.20	9.1–29.0		
2014 09 17	56917	2.55	6.8–38.4	SS	STEREO$_{post}$
2014 09 18	56918	1.29	6.3–26.5		
2014 09 20	56920	2.38	6.0–38.0		
2014 09 23	56923	3.00	6.0–39.0		
2014 09 24	56924	3.26	6.6–37.5		
2014 09 25	56925	1.81	6.2–39.0		

to that in which MAGIC previously reported evidence of emission (Albert et al. 2007). The X-ray spectral states were defined by using public *Swift*-BAT (15–50 keV) and *RXTE*-ASM (1.5–12 keV) data, except for the data taken in 2014 where only *Swift*-BAT was considered (since *RXTE*-ASM ceased science operations on the 3rd of January 2012). Between July and November 2007, the criteria used to prompt observations were a *Swift*-BAT flux larger than 0.2 counts cm^{-2} s^{-1} and a ratio between *RXTE*-ASM one-day average (in counts s^{-1} in a *Shadow Scanning Camera*) and *Swift*-BAT lower than 200. This criterion is in agreement with the one set by Grinberg et al. (2013) to define the X-ray states of Cygnus X-1 using *Swift*-BAT data and utilized also in the *Fermi*-LAT analysis (see Sect. 4.2.3): above 0.09 counts cm^{-2} s^{-1} (in the energy range between 15–50 keV) the microquasar stays in the HS+IS and below in the SS. The trigger criterion we selected is higher to achieve a count rate similar to that of the previous MAGIC hint. In July 2008, on top of the HS triggering criteria described above, observations were intensified following the X-ray superorbital modulation. The observations were triggered when the source

Table 4.4 UL to the integral flux above 200 GeV at 95% C.L. assuming a power-law spectrum with different photon indices, Γ. Credit: Ahnen et al. (2017)

Γ	Flux UL at 95% C.L. [$\times 10^{-12}$cm^{-2} s^{-1}]
2.0	2.20
2.6	2.44
3.2	2.62
3.8	2.71

was on the same superorbital phase as during the hint. Between June and October 2009, a new hardness ratio constraint using *RXTE*-ASM data of the energy ranges 5–12 keV and 1.3–2 keV was included: the observations were only stopped after 5 consecutive days of this ratio being lower than 1.2, to avoid interrupting the observations during the IS. In May 2011, the source was observed two nights based on internal analysis of public *Fermi*-LAT data that showed a hint at HE during a hard X-ray activity period. Since all the above mentioned data were taken during the HS, for completeness, Cygnus X-1 was observed in its SS on September 2014 to search for gamma-ray emission in this state at the same flux level as in the previous one. To define the X-ray state of the source, *Swift*-BAT public data was again used following Grinberg et al. (2013) criteria.

4.3.2   Search for Steady Emission

We search for steady VHE gamma-ray emission from Cygnus X-1 at energies greater than 200 GeV making use of the entire data set of almost 100 h. We did not find any significant excess and hence, we computed ULs assuming a simple power-law spectrum with different photon indices. The lower value, $\Gamma = 2$, is consistent with the results obtained in the HE band (see Sect. 4.2.3), while the upper one, $\Gamma = 3.8$, is constrained by the former MAGIC results ($\Gamma = 3.2 \pm 0.6$, Albert et al. 2007). Deviations in the photon index do not critically affect our results, quoted in Table 4.4, so all ULs obtained during this analysis are given at a 95% C.L. with $\Gamma = 3.2$, which is the photon index obtained for the hint of signal reported by the MAGIC Collaboration (Albert et al. 2007). For steady emission, we obtained an integral flux UL for energies greater than 200 GeV of 2.6×10^{-12} cm^2 s^{-1}. Differential flux ULs for the entire data sample can be found in Table 4.5.

4.3.2.1   Results During Hard State

Most of the observations ($\sim$83 h) of Cygnus X-1 were focused on the HS of the source. Observations under this X-ray spectral state were carried out between July 2007 and May 2011. However, no relevant excess can be reported during this spectral

Table 4.5 Differential flux ULs at 95% CL for the overall data sample assuming a power-law spectrum with photon index of $\Gamma = 3.2$. Credit: Ahnen et al. (2017)

Energy range	Significance	Differential flux UL for $\Gamma = 3.2$
[GeV]	[σ]	[$\times 10^{-13}$ TeV^{-1}cm^{-2} s^{-1}]
186–332	2.15	0.02
332–589	−0.14	0.33
589–1048	0.44	0.18
1048–1864	0.17	6.41
1864–3315	0.03	75.64

Table 4.6 Differential flux ULs at 95% C.L. for each X-ray spectral state. Credit: Ahnen et al. (2017)

Spectral state	Energy range	Significance	Differential flux UL for $\Gamma = 3.2$
	[GeV]	[σ]	[$\times 10^{-12}$ TeV^{-1} cm^{-2} s^{-1}]
HS	186–332	−2.57	0.20
	332–589	−0.03	3.70
	589–1048	2.09	1.31
	1048–1864	0.02	99.22
	1864–3315	0.51	16.34
SS	186–332	1.14	0.49
	332–589	1.22	0.11
	589–1048	0.06	4.71
	1048–1864	−1.23	51.62
	1864–3315	−1.34	16.37

state. The integral flux UL for energies greater than 200 GeV is 2.6×10^{-12} cm^{-2} s^{-1}. Differential flux ULs are listed in the upper part of Table 4.6.

In order to search for VHE orbital modulation, we carried out an orbital phase-folded analysis. To accomplish a good compromise between orbital phase resolution and significant amount of data, the binning in this analysis was 0.2. Moreover, in order to obtain enough statistics and cover the superior conjunction of the BH (phases 0.9–0.1), we started to bin the data at phase 0.1. No VHE orbital modulation is evident either. Integral ULs for this phase-folded analysis are shown in Table 4.7.

4.3.2.2 Results During Soft State

Cygnus X-1 was observed for a total of $\sim$14 h in the SS, bringing forth a clear difference on effective time with respect to the HS. Nevertheless, this corresponds to

Table 4.7 Orbital phase-wise 95% C.L. integral flux ULs for energies >200 GeV for HS and SS observations. The latter did not cover phases from 0.9–0.1, so no ULs are provided. Credit: Ahnen et al. (2017)

Spectral state	Phase	Eff. Time	Significance	Integral flux UL for $\Gamma = 3.2$
		[h]	[σ]	[$\times 10^{-12} cm^{-2} s^{-1}$]
HS	0.1–0.3	15.47	−0.77	7.89
	0.3–0.5	22.34	1.88	6.91
	0.5–0.7	14.08	0.00	21.32
	0.7–0.9	14.81	0.99	6.92
	0.9–0.1	15.62	−0.96	4.34
SS	0.1–0.3	2.58	0.45	19.32
	0.3–0.5	4.35	−1.23	7.96
	0.5–0.7	3.91	0.59	15.49
	0.7–0.9	3.64	0.23	18.23
	0.9–0.1	–	–	–

the post-upgrade period, in which MAGIC achieved its best sensitivity, $0.66 \pm 0.03\%$ of the Crab Nebula flux above 220 GeV in 50 h (Aleksić et al. 2016b), implying that the previous observations flux was nearly reached in only about 9 h. This data set guarantees, in turn, a full coverage of the X-ray spectral states that the source exhibits. Although steady gamma-ray emission in the SS, when no persistent jets are present, is not theoretically predicted, transient jet emission cannot be dismissed during this state, as it happens in the case of Cygnus X-3 (Tavani et al. 2009; Fermi LAT Collaboration et al. 2009). Nevertheless, we did not find any significant VHE gamma-ray emission from Cygnus X-1 in its spectrally soft state. Integral UL for energies beyond 200 GeV and $\Gamma = 3.2$ was set to $1.0 \times 10^{-11} cm^{-2} s^{-1}$. Differential ULs are quoted in the lower part of Table 4.6. The orbital phase-folded analysis did not yield any significant emission either. The integral ULs for this phase-folded study are also given in Table 4.7.

4.3.3 Search for Variable Emission

As done for the *Fermi*-LAT data analysis, because of the rapid flux variation spotted in Cygnus X-1, I carried out a daily-basis analysis for the 53 nights. This search yielded no significant excess in any of the nights and therefore, integral ULs (95% C.L.) for energies above 200 GeV were computed for single-night observations (listed in Table 4.8).

MAGIC results are included in the top panel of the multiwavelength lightcurve presented in Fig. 4.7. Along with MAGIC ULs, the figure shows data in the HE gamma-ray regime from *Fermi*-LAT (0.1–20 GeV) data analysis performed in this thesis, hard X-ray (*Swift*/BAT in 15–50 keV, Krimm et al. 2013), intermediate-soft X-ray (*MAXI* between 2–20 keV, Matsuoka et al. 2009), soft X-ray (quick-look results provided by the *RXTE*/ASM team in 3–5 keV) and radio data (AMI at 15 GHz and RATAN-600 at 4.6 GHz). The three transient episodes observed by *AGILE* are also marked. The X-ray states are identified in the lightcurve by the horizontal dashed line drawn in the X-ray pad at the level of 0.09 counts cm^{-2} s^{-1} (using *Swift*-BAT, Grinberg et al. 2013) and by the colored bands.

During this multi-year campaign, Cygnus X-1 did not display any X-ray flare like that in which the previous MAGIC 4.1σ result was obtained (Albert et al. 2007). This prevented us to observe the source under strictly the same conditions: the maximum *Swift*-BAT flux simultaneous to our observations happened on MJD 54379 (1.13σ, around superior conjunction of the BH) at the level of 0.23 counts cm^{-2} s^{-1}, close but still lower than 0.31 counts cm^{-2} s^{-1} peak around the *MAGIC hint*. However, we observed the microquasar in coincidence with the first *AGILE* flare. This transient episode seen by *AGILE* (on the 16th of October 2009, MJD 55120) showed a TS = 28.09 (4σ after trials) between 0.1–3 GeV with a gamma-ray flux of $(2.32 \pm 0.66) \times 10^{-6}$ photons cm^{-2} s^{-1} (Sabatini et al. 2010), which took place during the X-ray HS of Cygnus X-1. The corresponding MAGIC integral flux UL above 200 GeV for this day is 1.3×10^{-11} cm^{-2} s^{-1} (see Table 4.8). It is worth noting that, in accordance with our non-detection, the analysis of *Fermi*-LAT data did not show any significant signal (between 100 MeV and 20 GeV) on or around this date.

4.3.4 Discussion

MAGIC observations carried out between July 2007 and October 2014 for a total of ~100 h covered the two principal X-ray states of Cygnus X-1 with the main focus on the HS. We did not detect any significant excess from either all the data or any of the samples, including orbital phase-folded and daily analysis. This long-term campaign provided, for the first time, constraining ULs on the VHE emission of Cygnus X-1 at the two main X-ray states, the HS and the SS, separately as well as in an orbital binning base, which showed no hint of gamma-ray orbital modulation. This was possible thanks to a comprehensive trigger strategy that allowed to observe the source under flaring activity. The chosen photon index ($\Gamma = 3.2$ in this thesis, Crab-like in the previous MAGIC observations, Albert et al. 2007) and the addition of 30% systematic uncertainties contributed to obtain more robust ULs compared to the formerly ones reported by MAGIC.

The total power emitted by the jets during the HS in Cygnus X-1 is expected to be 10^{36}–10^{37} erg s^{-1} (Gallo et al. 2005). The integral UL 2.6×10^{-12} photons cm^{-2} s^{-1}, for energies greater than 200 GeV, obtained in this thesis with MAGIC corresponds to a luminosity of 6.4×10^{32} erg s^{-1} assuming a distance of 1.86 kpc (Reid et al. 2011). Therefore, the UL on the conversion efficiency of jet power to VHE gamma ray luminosity is 0.006–0.06%, similar to the one obtained previously

Table 4.8 *From left to right*: Date of the beginning of the observations in calendar and in Modified Julian Day (MJD), effective time after quality cuts, significance for an energy threshold of ∼150 GeV for *mono* observations (only MAGIC I) and ∼100 GeV for *stereoscopic* observations (separated by the horizontal line) and integral flux ULs at 95% C.L. for energies above 200 GeV computed on a daily basis. MJD 54656, 54657 and 54658 were analyzed separately according to each observational mode (see Table 4.3). Due to low statistics, neither the integral UL for MJD 55017 nor the significant for MJD 55116 were computed. Credit: Ahnen et al. (2017)

Date		Eff. time	Significance	Flux UL for Γ=3.2
[yyyy mm dd]	[MJD]	[h]	[σ]	[$\times 10^{-11}$ photons cm^{-2} s^{-1}]
2007 07 13	54294	1.78	−0.67	2.19
2007 09 19	54362	0.71	1.10	7.10
2007 09 20	54363	1.43	1.99	4.59
2007 10 05	54378	0.85	−0.84	1.84
2007 10 06	54379	1.85	0.02	1.21
2007 10 08	54381	1.95	0.99	2.88
2007 10 09	54382	0.77	−0.57	2.38
2007 10 10	54383	2.26	−0.04	1.05
2007 10 11	54384	0.76	1.68	2.26
2007 11 05	54409	0.58	0.31	4.38
2007 11 06	54410	0.96	−1.24	0.97
2008 07 02	54649	4.24	2.33	0.21
2008 07 03	54650	3.26	1.53	0.15
2008 07 04	54651	4.27	2.36	0.23
2008 07 05	54652	4.15	2.97	0.22
2008 07 06	54653	3.75	1.75	0.39
2008 07 07	54654	3.69	2.74	0.24
2008 07 08	54655	3.94	2.01	0.18
2008 07 09	54656	3.06	1.66	0.49
2008 07 10	54657	2.89	1.75	0.38
2008 07 11	54658	1.18	0.32	0.93
2008 07 09	54656	0.33	0.06	4.84
2008 07 10	54657	0.39	−1.22	3.11
2008 07 11	54658	0.32	1.83	8.81
2008 07 12	54659	2.51	0.11	1.16
2008 07 24	54671	0.62	−1.45	1.90
2008 07 25	54672	0.63	−0.15	2.30
2008 07 26	54673	0.84	−1.33	2.40
2008 07 27	54674	0.30	2.09	2.44
2009 06 30	55012	3.50	0.76	3.46
2009 07 01	55013	2.63	0.73	2.50

(continued)

Table 4.8 (continued)

Date		Eff. time	Significance	Flux UL for $\Gamma=3.2$
[yyyy mm dd]	[MJD]	[h]	[σ]	[$\times 10^{-11}$ photons cm^{-2} s^{-1}]
2009 07 02	55014	1.83	0.14	1.36
2009 07 05	55017	0.22	0.37	–
2009 10 08	55112	0.26	−1.85	1.11
2009 10 10	55114	0.67	0.19	1.50
2009 10 11	55115	2.03	0.32	3.10
2009 10 12	55116	2.34	–	2.19
2009 10 13	55117	0.95	1.53	3.87
2009 10 14	55118	1.98	−0.30	2.44
2009 10 16	55120	1.37	−2.99	1.30
2009 10 17	55121	0.96	−0.77	4.25
2009 10 18	55122	1.60	−0.27	3.05
2009 10 19	55123	0.68	−0.44	3.42
2009 10 21	55125	1.99	−1.90	1.09
2009 11 06	55141	0.37	−3.04	2.23
2009 11 07	55142	0.64	0.13	2.35
2009 11 13	55148	0.89	−1.23	3.06
2010 03 26	55281	0.78	1.75	10.92
2011 05 12	55693	1.35	0.09	1.38
2011 05 13	55694	1.20	−1.54	0.53
2014 09 17	56917	2.55	0.32	2.56
2014 09 18	56918	1.29	−0.99	1.25
2014 09 20	56920	2.38	0.08	2.13
2014 09 23	56923	3.00	0.85	2.85
2014 09 24	56924	3.26	−0.61	2.73
2014 09 25	56925	1.81	0.28	2.26

for Cygnus X-3 (Aleksić et al. 2010). Note that gamma-ray opacity in Cygnus X-3 is nevertheless about two orders of magnitude higher than in Cygnus X-1.

VHE emission from the jet large scale or jet-medium interaction regions above the sensitivity level of MAGIC can be ruled out, as these regions are not affected by gamma-ray absorption. On the binary scales, however, the non-detection is less conclusive because of pair creation in the stellar photon field. Models do predict VHE radiation as long as particle acceleration is efficient (e.g. (Pepe et al. 2015)). Formally, particle acceleration up to ~TeV energies can be reached in the jet on the binary region (Khangulyan et al. 2008), and thus 100 GeV IC photons should be produced, but this emission may be right below the detection level of MAGIC (as in

Zdziarski et al. 2016a), Fig. 6) even under negligible gamma-ray absorption. It could otherwise be that non-thermal particles cannot reach VHE IC emitting energies in the jet of Cygnus X-1. Besides inefficient acceleration, a very high magnetic field could also prevent particles to reach VHE, and even if these particles were present, a strong magnetic field can suppress intensely VHE photon production.

Nevertheless, one cannot dismiss the possibility of a transient emission as the one hinted by MAGIC in 2006. This flare took place during an orbital phase around the superior conjunction of the BH, where the gamma-ray absorption is expected to be the highest. The attenuation constraint may have been relaxed by an emitter at some distance from the BH (Albert et al. 2007), with its intrinsic variability possibly related for instance to jet-stellar wind interaction (Perucho et al. 2008; Owocki et al. 2009). On the other hand, even considering absorption by stellar photons, emission closer to the BH would be possible accounting for extended pair cascades under a reasonable intrinsic gamma-ray luminosity, although rather low magnetic fields in the stellar wind would be required (Zdziarski et al. 2009; see also Bosch-Ramon et al. 2008). Cygnus X-3, the other microquasar firmly established as a GeV emitter (Tavani et al. 2009; Fermi LAT Collaboration

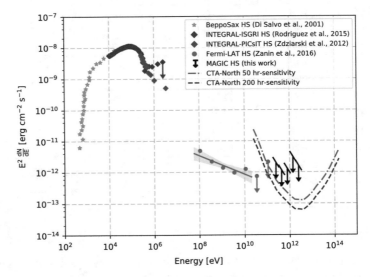

Fig. 4.8 SED of Cygnus X-1 covering X-ray, HE and VHE gamma-ray regimes during the HS. *BeppoSAX* soft X-ray data (in the keV band, green stars) is taken from Di Salvo et al. (2001), while for the hard X-ray band data from both *INTEGRAL*-ISGRI (10 keV–2 MeV, red diamond and UL; Rodriguez et al. 2015) and *INTEGRAL*-PICsIT (150 keV–10 MeV, brown diamond; Zdziarski et al. 2012) are displayed, given their incompatibility spectral results above 1 MeV. In the HE gamma-ray band (60 MeV-few hundred GeV, violet circles and ULs), results obtained with *Fermi*-LAT in HS are shown. At VHE, results from MAGIC during the HS are plotted (black) assuming a power-law function of $\Gamma = 3.2$. The dashed blue lines correspond to the 50 and scaled to 200 h sensitivity curves for CTA North. No statistical errors are drawn, except for the *Fermi*-LAT butterfly. Credit: Ahnen et al. (2017)

et al. 2009), displays a very different behavior from that of Cygnus X-1. The HE gamma-ray emission from Cygnus X-3 is transient, occurring sometimes during flaring activity of non-thermal radio emission from the jets (Corbel et al. 2012). If VHE radiation in microquasars were related to discrete radio-emitting-blobs with high Lorentz factor ($\Gamma \geq 2$), this may also happen in Cygnus X-1 during hard-to-soft transitions.

The multiwavelength emission from X-rays up to VHE gamma rays in Cygnus X-1 is shown in Fig. 4.8. The data used in this SED corresponds to the HS. The sensitivity curve for 50 and scaled to 200 h of observations with the future Cherenkov Telescope Array, CTA,[4] on the Northern hemisphere is showed along with the data. The spectral cutoff of the HE radiation from Cygnus X-1 is still unknown, although if the gamma-ray emission in the HS reaches $\sim$ TeV energies, the next generation of IACTs may be able to detect the system for long enough exposure times. Thus, to detect steady VHE emission from the jets, future more sensitive instruments, as CTA, would be needed. This instrument could provide valuable information of the VHE gamma-ray production in Cygnus X-1 (HE spectral cutoff, energetics, impact of gamma-ray absorption/IC cascades), as well as allow the study of possible short-term flux variability.

4.4 Conclusions

We obtained, for the first time, a high-significance detection of a BH binary system in the HE regime. We established the detection of a point-like LAT source spatially coincident with the microquasar Cygnus X-1 at the level of TS = 53 for energies above 60 MeV, using 7.5 years of **Pass 8** *Fermi*-LAT data. By analyzing the sample at the two different main X-ray states, HS and SS, we could confirm a correlation between GeV emission and hard X-rays: the source is only detected during the former state, i.e. when steady relativistic radio-emitting jets are displayed (TS = 49). During this spectrally hard state, the emission was detected around the superior conjunction of the compact object ($\phi > 0.75$ and $\phi < 0.25$, at TS = 31), while it becomes undetectable at the inferior conjunction, evidencing a hint of orbital modulation in gamma rays.

We could constrain the production site of HE photons at distances $Z > 10^{11}$ cm from the BH along the jet, to prevent pair production mechanism faints the GeV photons and account for the dominant stellar photon radiation, and $Z < 10^{13}$ cm, to confine the HE emission inside the binary system given the presumable orbital variability.

On the other hand, observations with the MAGIC telescopes allowed us to discard the interaction between the relativistic jets and the surrounding medium as production site for VHE gamma rays at the level of the MAGIC sensitivity, taken into account that this region is not affected by photon-photon absorption. Nevertheless, TeV gamma rays produced inside the binary still remains a valid possibility.

---

[4]Taken from https://www.cta-observatory.org/science/cta-performance/.

References

Acero F et al (2015) ApJS 218:23
Aharonian FA et al (1981) Ap&SS 79:321
Ahnen ML et al (2017) MNRAS 472:3474
Akharonian FA et al (1985) Ap&SS 115:201
Albert J et al (2007) ApJ 665:L51
Aleksić J et al (2010) ApJ 721:843
Aleksić J et al (2012a) J Cosmol Astropart Phys 10:032
Aleksić J et al (2012b) Astropart Phys 35:435
Aleksić J et al (2016a) Astropart Phys 72:61
Aleksić J et al (2016b) Astropart Phys 72:76
Aliu E et al (2009) Astropart Phys 30:293
Bodaghee A et al (2013) ApJ 775:98
Bolton CT (1972) Nature 235:271
Bosch-Ramon V et al (2008) A&A 489:L21
Brocksopp C et al (1999) Mon Not Roy Astron Soc 309:1063
Brocksopp C et al (1999) A&A 343:861
Bulgarelli A et al (2010) The Astronomer's Telegram, 2512
Collaboration FLAT et al (2009) Science 326:1512
Corbel S et al (2012) MNRAS 421:2947
Di Salvo T et al (2001) ApJ 547:1024
Esin AA et al (1998) ApJ 505:854
Fender RP et al (2004) MNRAS 355:1105
Fender RP et al (2006) Mon Not Roy Astron Soc 369:603
Fomin VP et al (1994) Astropart Phys 2:137
Gallo E et al (2005) Nature 436:819
Gies DR et al (2008) ApJ 678:1237
Gosnell NM et al (2012) ApJ 745:57
Grinberg V et al (2013) A&A 554:A88
Guenette R et al (2009). arXiv:0908.0714
Jackson JC (1972) Nat Phys Sci 236:39
Jourdain E et al (2012) ApJ 761:27
Khangulyan D et al (2008) MNRAS 383:467
Khangulyan D et al (2014) ApJ 783:100
Krimm HA et al (2013) ApJS 209:14
Lachowicz P et al (2006) MNRAS 368:1025
Laurent P et al (2011) In: X-ray astrophysics up to 511 keV
Malyshev D et al (2013a) MNRAS 434:2380
Malyshev D et al (2013b) MNRAS 434:2380
Malzac J et al (2008) A&A 492:527
Matsuoka M et al (2009) PASJ 61:999
McConnell ML et al (2002) ApJ 572:984
Munar-Adrover P et al (2016) ApJ 829:101
Nice DJ et al (2013) ApJ 772:50

Orosz JA et al (2011) ApJ 742:84
Owocki SP et al (2009) ApJ 696:690
Pepe C et al (2015) A&A 584:A95
Perucho M et al (2008) A&A 482:917
Poutanen J et al (2008) MNRAS 389:1427
Priedhorsky WC et al (1983) ApJ 270:233
Reid MJ et al (2011) ApJ 742:83
Rico J (2008) ApJ 683:L55
Rodriguez J et al (2015) ApJ 807:17
Romero GE et al (2014) A&A 562:L7
Rushton A et al (2012) MNRAS 419:3194
Russell DM et al (2010). ArXiv e-prints
Sabatini S et al (2010) ApJ 712:L10
Sabatini S et al (2013) ApJ 766:83
Stirling AM et al (2001) Mon Not Roy Astron Soc 327:1273
Szostek A et al (2007) MNRAS 375:793
Tavani M et al (2009) Nature 462:620
van Leeuwen F (2007) A&A 474:653
Wen L et al (1999) ApJ 525:968
Yoon D et al (2016) MNRAS 456:3638
Zanin R et al (2016) A&A 596:A55
Zdziarski AA et al (2016a). ArXiv e-prints
Zdziarski AA et al (2016b) MNRAS 455:1451
Zdziarski AA et al (2009) MNRAS 394:L41
Zdziarski AA et al (2011) MNRAS 412:1985
Zdziarski AA et al (2012) MNRAS 423:663
Zdziarski AA et al (2013) MNRAS 436:2950
Ziółkowski J (2014) MNRAS 440:L61

Chapter 5
Cygnus X-3

5.1 History

Cygnus X-3, discovered 50 years ago (Giacconi et al. 1967), is one of the brightest X-ray sources in our Galaxy and the strongest radio source among the X-ray binaries, displaying two-sided transient powerful relativistic radio jets (Martí et al. 2001; Miller-Jones et al. 2004). Located in the Galactic plane ($l = 71.32°$ and $b = +3.09°$), the system lies at a most likely distance of 7.4 ± 1.1 kpc from Earth, based on a Bok globule's emission study situated along the light of sight of Cygnus X-3 (McCollough et al. 2016). The features of this microquasar are not representative of the typical X-ray binary. First of all, despite hosting a WR star (van Kerkwijk et al. 1992), it shows an orbital period of 4.8 h, more common of low-mass binaries. This value was ascertained by X-ray and IR flux orbital modulation detected during flaring activity of the source (Parsignault et al. 1972; Becklin et al. 1973). This periodic intensity variations may arise from the scattering or absorption of the emission from the stellar wind, as in other binary systems. On the other hand, the proximity between the compact object and the companion star gives rise to an unusual strong absorption, as a consequence of the embed of the system within the companion stellar wind. This complicates the understanding of the system, from which the nature of the compact object is unclear so far: it can be composed by a 1.4 $M_\odot$ NS (Stark et al. 2003) or a $< 10 M_\odot$ BH (Hanson et al. 2000). Nevertheless, the latter scenario involving a BH is reinforced given that the only other known binaries composed by WR stars host this type of compact object (see Prestwich et al. 2007 for IC 10 X-1 and Carpano et al. 2007 for NGC 300 X-1) and the resemblances that the Cygnus X-3 spectral features present with respect to the spectra of BH transients.

Actually, despite the difficult task that the study of the Cygnus X-3 spectrum implies, given the strong X-ray absorption, the two main X-ray spectral states typical from BH transients are observed (the HS and the SS). The HS was studied by Hjalmarsdotter et al. (2008), which determined an unusually low energy cutoff of ~ 20 keV (compared to the typical cutoff at hundred keV displayed by the counterparts).

© Springer Nature Switzerland AG 2018
A. Fernández Barral, *Extreme Particle Acceleration in Microquasar Jets and Pulsar Wind Nebulae with the MAGIC Telescopes*, Springer Theses,
https://doi.org/10.1007/978-3-319-97538-2_5

A possible explanation for this additional peculiarity could be the presence of a very massive BH with >20 $M_\odot$ forming the system. With such mass, the last stable orbit would be further away from the compact object than in other sources and hence, Cygnus X-3 could not reach temperatures as high as other binaries. Thus, the HS needs to be modeled by the Comptonization of hybrid (thermal and non-thermal) electron population along with a Compton reflection component, instead of only thermal one as expected in the standard scenario.

Although Cygnus X-3 is always detected in the radio band, its flux level in this wavelength can vary several orders of magnitude during its frequent radio outbursts, the first of which was reported at the beginning of the 70's by Gregory et al. (1972). Waltman et al. (1994, 1995, 1996) discussed the existence of four radio states: *quiescent state*, with fluxes around ∼50–200 mJy that lasts months; *minor flaring episodes* with ≲300 mJy; *quenched periods* in which the radio flux is as low as ≲30 mJy; and *major flaring periods*, during which the flux level increases large fraction up to ∼1–20 Jy. These strong radio flares happen during the SS. The correlation between the radio and the soft X-rays was deeply studied afterwards by Szostek et al. (2008). This connection can be split into six states, following the previous division, represented in the so-called *saxophone plot* (see Fig. 5.1), which I summarize below:

- **Quiescence state**: Low flux in both energy bands, radio and soft X-rays. Variation is positively correlated.
- **Minor flaring state**: State in which small oscillations can happen as the source is overpassing the X-ray transitional level (set to 3 counts/s using *RXTE*-ASM data in the 3–5 keV band). To the left of this level, the source stays in the HS, at the right it is in the SS. The radio level remains below its transitional level of 300 mJy.
- **Suppressed state**: There is anti-correlation between fluxes. While the source is finally entering in the SS in which flux of soft X-rays increases, the radio flux keeps decreasing. It is not always followed by a major flare.
- **Quenched state**: The radio flux is at its lowest level, while the soft X-rays show one of the highest fluxes. This state is directly followed by a radio outburst.
- **Major flaring state**: The soft X-rays keep roughly constant, whilst radio suffers huge variation of several factors of magnitude.
- **Post flaring state**: It corresponds to a returning state to either the minor flaring or the suppressed state.

Cygnus X-3 was detected in the HE regime (above 100 MeV) by both *AGILE* and *Fermi* (Tavani et al. 2009b; Fermi LAT Collaboration et al. 2009b). *AGILE* detected four gamma-ray flares of 1–2 days each one between mid-2007 and mid-2009, leading to an average spectrum between 100 MeV and 3 GeV described by a power-law function with photon index 1.8 ± 0.2. On the other hand, *Fermi* detection took place mainly during two activity periods of around two months (MJD 54750–54820 and MJD 54990–55045), an estimated spectrum was as well defined by a power-law distribution but with a softer index of $2.70 \pm 0.05_{stat} \pm 0.20_{syst}$. All GeV detections happened during the SS of the source and before a radio major flare. *Fermi* estimated this gap between HE and radio photons on (5 ± 7) days. These results evidence the episodic nature of the HE emission, opposite to the steady radiation detected

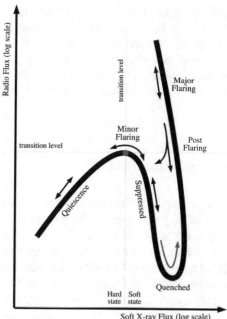

Fig. 5.1 Cygnus X-3 *saxophone plot* that represents the evolution of the source through the different radio and X-ray states. The dotted lines correspond to the transition level (determined at 0.3 Jy and 3 counts/s with GBI radio and *RXTE*-ASM soft X-rays data, respectively). The six different defined states are labeled along the path in which arrows show the possible direction that the system can follow. Modified plot from Szostek et al. (2008)

Table 5.1 Cygnus X-3 GeV flaring periods determined making use of the UB *Fermi*-LAT pipeline results

Name	Period [MJD]	Status
Flare 1	54750–54820	Reported in Fermi LAT Collaboration et al. (2009b)
Flare 2	54990–55045	
Flare 3	55585–55610	Reported in Corbel et al. (2012)
Flare 4	55640.5–55643.5	
Flare 5	57398–57415	Unpublished
Flare 6	57622–57653	Reported in this thesis at VHE
Flare 7	57799–57873	Unpublished

in Cygnus X-1 (see Chap. 4). After these two flaring activity periods that yielded to its detection in the HE regime, Cygnus X-3 underwent another five outburst (see Table 5.1).

This microquasar was also claimed to be a TeV emitter (see e.g. Vladimirsky et al. 1973) and even a PeV emitter (see Bhat et al. 1986). Nevertheless, this presumable detection in the TeV regime was not confirmed by more recent and sensitive instruments, as MAGIC (Aleksić et al. 2010b) or VERITAS (Archambault et al. 2013). Former MAGIC observations amounted a total of ∼60 h of good quality data, between 2006 and 2009 in stand-alone mode with MAGIC I. Observations during this previous campaign were triggered by either radio or gamma-rays alerts sent by RATAN-600 and *AGILE*, respectively. These MAGIC observations yielded to an integral UL at 95% CL of 2.2×10^{-12} photons cm^{-2} s^{-1} at energies greater than 250 GeV.

5.2 Observations and Data Analysis

Cygnus X-3 observations belong to a long-term observational campaign within the MAGIC internally so-called Key Observation Program (KoP). These projects comprehend those sources that, due to the high scientific interest and impact, have observational priority. Sources under KoP are not usually scheduled commonly, but their observations are triggered based on multiwavelength information in order to observe them at the most appropriate period. In the case of Cygnus X-3, the trigger criterion is based on a *Fermi* pipeline performed at the Universitat de Barcelona (UB). Currently, we observe the system when the TS on their daily analysis of public *Fermi* data (using photon-like events between 00:00:01 and 00:00:01 of two consecutive days, approximately) is higher than 13. In order to decide to keep or interrupt observations, I analyze the MAGIC data as soon as it is available the following day: if a significant hint is obtained in this analysis or the *Fermi* pipeline fulfills the aforementioned criterion again, observations are scheduled for the next day, otherwise they are interrupted. The Cygnus X-3 KoP started on November 2013 and keeps active up to now. The source is observable from April to November, due to its position in the Cygnus Constellation, for a broad Zd range of (12–50)°.

Cygnus X-3 is observed making use of the *wobble* tracking mode (see Sect. 2.4.2.1) but with non-standard positions in order to avoid the speculative pulsar binary TeV J2032+4130 and the binary star systems WR 146 and WR 147, situated in the FoV of the target microquasar. Thus, the *wobble* positions for Cygnus X-3 observations are placed at a RA angles of 51°, 141°, 231° and 321°, at a nominal offset distance of 0.4° (see Fig. 5.4).

For the latest observational period (August–September 2016 campaign), although most of the data were taken under dark conditions, MAGIC pointed to Cygnus X-3 under different moonlight levels in order to extend observations as much as possible. Thus, a proper analysis with dedicated MC and OFF data was applied in each case. The data was split into six categories according to the NSB background level, which are: $1 \times$ NSB$_{dark}$ (corresponding to *dark* conditions), $2 - 3 \times$ NSB$_{dark}$, $3 - 5 \times$ NSB$_{dark}$, $5 - 8 \times$ NSB$_{dark}$, $8 - 12 \times$ NSB$_{dark}$ and $12 - 18 \times$ NSB$_{dark}$. It is worth mentioning that all moon data were taken using nominal HV (∼1.25 kV), including the highest levels, without resorting to the reduced HV. Therefore, I summarize in Table 5.2 the cleaning levels and *size* cuts applied for the Cygnus X-3 analysis as well as the pedestal mean and RMS distribution values used in the additional noise included in the MC and OFF data.

Table 5.2 Image cleaning levels for the Cygnus X-3 analysis

NSB [× NSB$_{dark}$]	Cleaning levels Lvl$_1$:Lvl$_2$ [phe]	Pedestal distribution [phe]	RMS distribution [phe]	Size cut [phe]
1–2	6:3.5	–	–	50
2–3	7:4.5	3.0	1.3	80
3–5	8:5	3.6	1.5	110
5–8	9:5.5	4.2	1.7	150
8–12	10:6	4.8	2.0	210
12–18	13:8	5.8	2.3	250

5.2.1 August–September 2016 Flare

During summer 2016, Cygnus X-3 underwent a flaring activity in both radio and HE gamma-ray regimes. The outburst period was initiated the 21st of August (MJD = 57621), reference based on the UB *Fermi*-LAT pipeline, which showed a TS = 20.5 ($\sim 4.5\sigma$) that day. This enhanced activity continued almost a month, during which Cygnus X-3 was firmly detected in three occasions by this *Fermi*-LAT pipeline: on MJD 57631 with TS $\sim$ 28 ($\sim 5.3\sigma$), on MJD 57647 with TS $\sim$ 33 ($\sim 5.7\sigma$) and on MJD 57649, when a major flare with TS $\sim$ 73 ($\sim 8.5\sigma$) happened. Although no other days showed significance well above 5σ, several intervening days presented TS higher than the internal criterion of TS = 13. The TS light curve obtained from the daily *Fermi*-LAT analysis performed at UB since August 2008 is depicted in Fig. 5.2. This HE emission during August–September 2016 was confirmed by *AGILE* (Piano et al. 2016) and the *Fermi*-LAT Collaboration (Cheung and Loh 2016) as well.

On the other hand, this HE activity period started while the radio emission began to decrease, revealing that Cygnus X-3 was entering in the SS (Trushkin et al. 2016a). Actually, radio flux at 4.6, 8.2 and 11.2 GHz from RATAN-600 reached *quenched* level (around 10–30 mJy) during MJD 57622–57625, foreshadowing a major flare. Thus, on the 13th of September (MJD 57643.8) a giant flare of 15 Jy at 4.6 GHz was detected by RATAN-600 (Trushkin et al. 2016b). The combination of all these conditions provided an excellent opportunity to observe the source under the state in which VHE gamma rays are expected.

MAGIC follow-up observations started on the 23rd of August, triggered by the internal UB *Fermi*-LAT pipeline. The campaign was extended up to the 22nd of September, point at which MAGIC had observed the system for 18 nights (among which only one was discarded afterwards during the analysis for presenting *cloudiness* above 40%). From the 27th of August to the 7th of September, observations were carried out uninterruptedly every night aiming to observe the source the greatest possible time. This deep MAGIC campaign allowed us to observed the source for a total of $\sim$70 h of good quality data and, most important, from the very beginning of the HE emission until it ceased. Therefore, for the first time, we were able to obtain

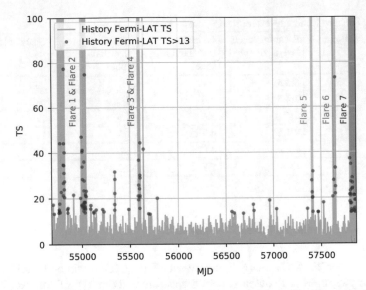

Fig. 5.2 Cygnus X-3 TS evolution taken from the UB *Fermi*-LAT pipeline. Days with TS higher than 13 are marked with a dark blue dot. All HE flaring periods of the source are highlighted and labeled according to Table 5.1. A zoom view of Flare 6 reported in this thesis can be found in Fig. 5.6

a great amount of data from Cygnus X-3 during an entire outburst that behaved as the one that led to its detection in the HE gamma-ray regime in 2009 (Fermi LAT Collaboration et al. 2009b).

Figure 5.3 shows a schematic view of the *saxophone plot* given by Szostek et al. (2008), in which MAGIC observations are set in context. Given the lack of simultaneous soft X-ray data that covers the entire observational period, I defined two periods based on 15 GHz Owens Valley Radio Observatory (OVRO) radio information (provided under private communication): *Period 1*, from MJD 57623 to 57638, in which Cygnus X-3 was most of the time below the radio transition level (with an upper value of ~500 mJy); and *Period 2*, from MJD 57650 to 57653, during which the source was entirely at a *major flaring* state (lower value at ~3 Jy and reaching around 9.5 Jy).

5.3   Results Before Flare 2016

From the beginning of the project until August 2016, we observed Cygnus X-3 for 16 nights, all of them under dark conditions. The dates and the corresponding significance (for FR and LE cuts, see Sect. 2.4.3.9) are listed in Table 5.3. No significant excess was found in any of these nights. The only remarkable episode happened on the 29th of November 2015, when a hotspot at the level of 5σ appeared close to the nominal position at a distance of ~0.23° (approximated coordinates RA = 308.03°,

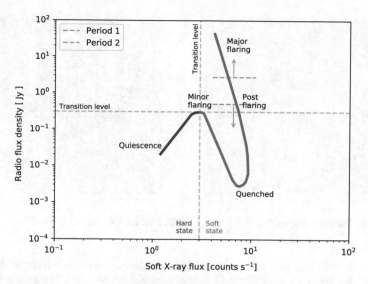

Fig. 5.3 Modified *saxophone plot* from Szostek et al. (2008). Due to the lack of soft X-ray information, MAGIC observations periods are marked with two arrows based on OVRO data: the green one corresponds to *Period 1* (MJD 57623–57638) and the orange one to *Period 2* (MJD 57650–57653)

Table 5.3 Cygnus X-3 observations from November 2013 until August 2016. *From left to right*: Calendar date, effective time after quality cuts, significance with FR cuts and LE cuts at the nominal position

Date	Effective time	Significance	
[yyyy mm dd]	[hr]	FR	LE
		[σ]	[σ]
2013 11 26	1.30	1.18	−0.48
2014 05 05	0.86	0.32	−1.24
2014 05 06	2.10	0.32	0.24
2014 05 25	2.10	−1.67	−0.15
2014 05 26	2.15	1.76	0.43
2014 06 09	1.32	0.05	0.76
2014 10 11	0.98	1.68	2.33
2014 1012	2.62	1.60	−0.50
2014 10 14	2.29	−0.09	−1.50
2014 10 16	0.53	−0.97	−1.46
2014 10 28	0.16	0.76	−1.06
2015 07 29	5.79	−3.10	−0.96
2015 11 11	0.65	0.51	−0.56
2015 11 29	0.88	−1.56	0.55
2015 11 30	0.56	0.41	−0.14
2016 05 17	0.96	0.0	1.07

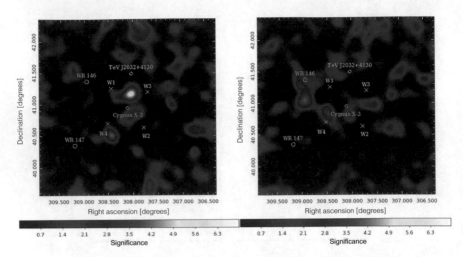

Fig. 5.4 Cygnus X-3 LE skymaps for the 29th (*left*) and the 30th (*right*) of November 2015. The first night a hotspot at a distance of $\sim$0.23° from Cygnus X-3 (green diamond) pop-up at the level of 5σ, not coincident with any known TeV source or candidate in the FoV. This putative excess disappeared the following night. The source TeV J2032+4130 (yellow diamond) and the binary star systems WR 146 and WR 147 (cyan circles), as well as the wobble positions from the Cygnus X-3 observations (white crosses), are highlighted and labeled

Dec $= 41.18°$). This excess was not coincident with the location of the source TeV J2032+4130 either (see Fig. 5.4). The significance, computed with `Odie` following the Off from Wobble Partner (OfWP) (see detailed information of this background estimation in Sect. 8.2), was however $\sim$2.5σ. Nevertheless, and even though the signal from Cygnus X-3 was compatible with background from both θ^2 plots and skymaps, observations were extended the 30th of November triggered by this interesting hotspot in the FoV. On the 30th, the skymap did not present any highlighted region and observations were stopped. It is worth highlighting that no excess on that particular position was reported so far at any other wavelength.

5.4   Results of the August–September 2016 Flare

I searched for VHE gamma-ray emission in the entire period of activity, i.e. making use of the whole available data sample of $\sim$70 h. However, no excess was found, being the signal compatible with background with a significance of -1.27σ applying Eq. 2.12 with FR cuts. Thus, integral UL at 95% C.L. assuming a power-law distribution with photon index $\Gamma = 2.6$ for energies above 300 GeV was computed. This calculation yielded an UL of 2.6×10^{-13} photons cm^{-2}s^{-1}. Note that the minimum energy is constrained by the highest moonlight level data, to which a size cut of 250 phe is applied (see Table 5.2), implying an energy threshold of $\gtrsim$250 GeV. On the

Table 5.4 MAGIC integral ULs above 300 GeV at 95% C.L. assuming a power-law spectrum with different photon indices, Γ

Γ	Flux UL at 95% C.L. $[\times 10^{-13}$ photons cm^{-2}s$^{-1}]$
2.0	2.4
2.6	2.6
3.2	2.7

other side, variations of $\pm 30\%$ on the photon index were probed and gave rise to less than 10% difference (see Table 5.4). Therefore, for the rest of the discussion, I decided to use the standard $\Gamma = 2.6$, which allowed me to compare the results obtained in this thesis with the previous MAGIC ULs (Aleksić et al. 2010b), for which the computation a Crab-like spectrum was also assumed. In order to decrease the energy threshold up to 100 GeV and be able to extend results at lower energies, I also made use of the data sample taken under dark conditions only ($\sim$52 h) to compute ULs. The integral UL under dark conditions, assuming the above mentioned spectral shape, is 5.1×10^{-12} photons cm^{-2}s^{-1}. Differential ULs, for the overall data set and only dark sample, are listed in Table 5.5 and presented in Fig. 5.5 along with former MAGIC results as well as with the *Fermi*-LAT spectrum obtained by an independent analysis for the flaring period of 2016. The *Fermi*-LAT spectrum was computed using FRONT+BACK photons from the `Pass 8` data between MJD 57642–57652 that encompasses *Period 2*, in which the peak at TS = 73 was observed in the UB pipeline. In this analysis, the background model was created from 3FGL sources, including the pulsation emission from TeV J2032+4130 (for which radio ephemeris from private communication were used), as well as the corresponding diffuse emission (*gll_iem_v06* and *iso_P8R2_SOURCE_V6_v06*). The off-pulse emission from TeV J2032+4130 was described as a power-law with index $\Gamma = 2.54$ (value obtained for the 8-years analysis data) and the Cygnus Cocoon contribution was fixed by the value given in Ackermann et al. (2011). To avoid contamination from another nearby pulsar, PSR J2021+40 located at 2.3°, the analysis was performed above 300 MeV (energy at which the *Fermi*-LAT PSF is smaller than 2°). Thus, during this *Period 2*, GeV emission was detected at TS = 27 up to $\sim$20 GeV and it is well defined by $dN/dE = (1.05 \pm 0.59) \times 10^{-11} (E/1.3\,\text{GeV})^{3.23 \pm 0.56}\,\text{GeV}^{-1}\,\text{cm}^{-2}\,\text{s}^{-1}$. It is worth mentioning that no detection was achieved for *Period 1* (TS = 13). The implications of the new results are discussed in the next section.

On the other hand, as previously introduced, Cygnus X-3 is a highly variable source in X-rays and HE gamma rays, and hence, one cannot discard this type of variation in the VHE band. Therefore, I performed a daily basis analysis. Nevertheless, no significant excess or hint was found during this search. Integral ULs above 100 and 300 GeV (depending on the observational conditions) for each night are quoted in Table 5.6, with the respective significance and effective time. These results are set in context in Fig. 5.6 together with the TS from the UB *Fermi*-LAT pipeline, hard X-ray data from *Swift*-BAT (15–50 keV, Krimm et al. 2013) and soft X-ray data from MAXI (2–4 keV, Matsuoka et al. 2009). The low *Swift*-BAT flux during the entire observational campaign, along with the decrease in radio flux (which reached *quenched* state, Trushkin et al. 2016b) before major outbursts, evidences the SS in

Table 5.5 MAGIC differential flux ULs for Cygnus X-3 during the August–September 2016 flare, assuming a power-law spectrum with spectral index of $\Gamma = 2.6$

Energy range [GeV]	Differential flux ULs [TeV^{-1} $cm^{-2}s^{-1}$]	Observational conditions
75.4–119.4	2.7×10^{-10}	dark
119.4–189.3	6.8×10^{-11}	dark
189.3–300.0	1.1×10^{-11}	dark
300.0–475.5	1.1×10^{-12}	dark + moon
475.5–753.6	5.0×10^{-13}	dark + moon
753.6–1194.3	2.6×10^{-13}	dark + moon
1194.3–1892.9	1.5×10^{-13}	dark + moon
1892.9–3000	2.9×10^{-14}	dark + moon
3000.0–4754.7	9.7×10^{-15}	dark + moon

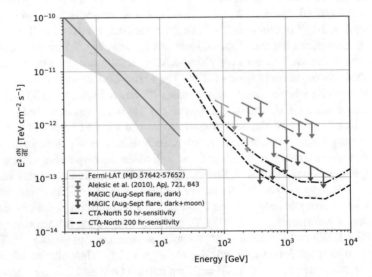

Fig. 5.5 Cygnus X-3 SED for the flaring period August–September 2016. The blue butterfly corresponds to the *Fermi*-LAT results obtained by an independent analysis of the HE data MJD 57642-57652. Former MAGIC results given by Aleksić et al. (2010b) are depicted in grey, while ULs obtained in this thesis are shown in orange. At lower energies, only dark data was used (~52 h, light orange) while the whole data sample was taken (~70 h, dark orange) above 300 GeV. Sensitivity curves of CTA-North for 50 h (dot-dashed line) and 200 h (dashed lines) of observations are shown too

which Cygnus X-3 stayed during the flaring activity. HE and radio outbursts happened almost simultaneously, separated approximately 2 days, in agreement within the errors with the results obtained by the *Fermi*-LAT Collaboration during the 2009 flare (5 ± 7 days, Fermi LAT Collaboration et al. 2009b). This gap is assumed by comparing the highest flux level in each regime: two days before the major

Table 5.6 Daily basis analysis for the flaring period between August–September 2016. *From left to right*: Date, in calendar and MJD, Zd range, effective time, significance applying FR and LE cuts, and integral ULs above 150 GeV (dark data only) and 300 GeV (dark+moon data). MJD 57623, 57650, 57651, 57652 and 57653 were observed under all aforementioned moonlight levels and hence, significance with LE cuts is not computed. MJD 57650 shows no integral UL above 150 GeV due to low statistics

Date		Zd	Eff. Time	Significance			Integral ULs (E>150 GeV)	Integral ULs (E>300 GeV)
				FR	LE			
[yyyy mm dd]	[MJD]	[°]	[hr]	[σ]	[σ]		[$\times 10^{-12}$ photons cm^{-2}s^{-1}]	
2016 08 23	57623	12–46	5.9	1.03	–		8.9	2.2
2016 08 24	57624	12–27	1.8	−1.63	0.08		3.9	–
2016 08 27	57627	22–32	1.0	1.25	0.83		9.0	–
2016 08 29	57629	20–30	1.0	−0.66	−0.35		10.8	–
2016 08 30	57630	20–30	1.0	0.21	−0.29		12.8	–
2016 08 31	57631	12–30	3.0	−0.35	−0.47		5.4	–
2016 09 01	57632	12–28	2.8	0.61	−0.97		7.4	–
2016 09 02	57633	10–50	5.7	0.12	0.35		3.4	–
2016 09 03	57634	10–50	5.8	−2.95	1.20		2.8	–
2016 09 04	57635	10–50	5.8	−0.26	−1.30		8.9	–
2016 09 05	57636	10–50	5.8	−1.53	−0.02		7.2	–
2016 09 06	57637	10–50	5.2	0.26	0.30		5.8	–
2016 09 07	57638	10–50	5.5	−0.54	−1.16		4.8	–
2016 09 19	57650	12–50	4.7	0.05	–		–	2.1
2016 09 20	57651	12–49	4.8	−0.68	–		10.1	1.0
2016 09 21	57652	12–50	5.0	0.88	–		5.6	2.7
2016 09 22	57653	10–50	3.7	−0.48	–		7.7	1.5

Fermi-LAT flare with TS = 73 on MJD 57649, radio flux increased to its maximum (see e.g. Trushkin et al. 2016c; Egron et al. 2016). Nevertheless, note that this presumable separation is got based on the results from the UB *Fermi*-LAT pipeline. For a more precise gap calculation between *Fermi*-LAT and radio flares, a dedicated *Fermi*-LAT flux light curve is needed.

The HE gamma-ray emission detected by *Fermi*-LAT (Fermi LAT Collaboration et al. 2009b) was found to be orbitally modulated, with its maximum coincident with the superior conjunction of the compact object, which corresponds to phase 0. I computed phase-folded analysis following the ephemeris of $T_0 = 2440949.892 \pm 0.001$ Julian Day (JD) given by Singh et al. (2002). Owing to the great amount of data and the short orbital period of 4.8 h, binning of 0.1 was possible. No hint was seen at any phase, including at the superior conjunction. The integral UL for each phase, with the corresponding FR significance and the effective time, is illustrated in Fig. 5.7.

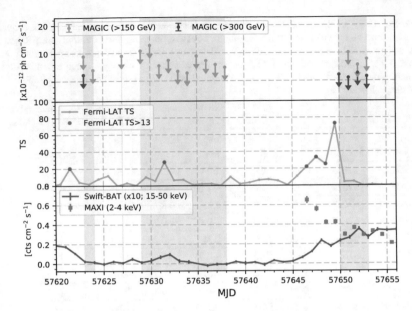

Fig. 5.6 Cygnus X-3 light curve for the flaring period August–September 2016. *From the top to the bottom*: MAGIC ULs obtained in this thesis for a minimum energy of 150 GeV (dark data only) and 300 GeV (dark+moon), TS evolution from the UB pipeline where days with TS > 13 are highlighted in dark blue and public available hard and soft X-rays from *Swift*-BAT and MAXI, respectively. Gray bands corresponds to the MAGIC observations. No integral UL above 150 GeV is shown for MJD 57650 because the total amount of dark data for this day corresponds to less than 5 min

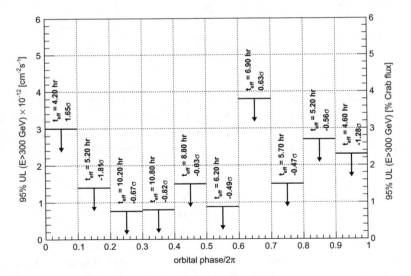

Fig. 5.7 Cygnus X-3 phasecogram assuming the ephemeris of $T_0 = 2440949.892 \pm 0.001$ (Singh et al. 2002). The corresponding significance and effective time for each bin are labeled

5.5 Discussion and Conclusions

As discussed in Sect. 3.2.4, VHE emission is postulated to happen inside the jets via both leptonic and hadronic mechanisms. This emission can occur steadily, as suggested for Cygnus X-1 in which gamma-ray emission originates in the persistent jet during the HS (see Chap. 4); or it could present transitional behavior, as expected in Cygnus X-3 from powerful radio-emitting blobs during the SS.

By the time of the beginning of this thesis, Cygnus X-3 was the best candidate among the microquasars to emit VHE, as it was the only source of this type from which HE gamma rays had been firmly detected (Fermi LAT Collaboration et al. 2009b; Tavani et al. 2009b). Therefore, MAGIC focused the efforts on observing this source during HE flares by a dedicated trigger system. Eventually, we were able to observe the source during an entire strong radio and HE outburst in 2016, from the very beginning of the flare activity until the peak of emission ceased a month later. Nevertheless, no detection was achieved at VHE for any of the analyzed data samples. The obtained integral UL for energies greater than 300 GeV was 2.6×10^{-13} photons $cm^{-2}s^{-1}$ and 5.1×10^{-12} photons $cm^{-2}s^{-1}$ above 100 GeV, corresponding to a luminosity UL of $L_\gamma(>300GeV) < 2.0 \times 10^{33}$ erg s^{-1} and $L_\gamma(>100GeV) = 1.3 \times 10^{34}$ erg s^{-1}, respectively, assuming a distance of $\sim$7 kpc. Given that the expected luminosity inside the relativistic jets is around 10^{37} erg s^{-1} (Martí et al. 2001), the UL on the conversion efficiency of the jet power to the VHE gamma-ray luminosity is smaller 0.02% ($>$300 GeV) and 0.13% ($>$100 GeV).

One has to consider the extremely high absorption seen in X-rays by the wind of the WR companion, which might be not negligible for VHE gamma rays either. In the case of Cygnus X-3, the WR star has a temperature of $T_* \sim 10^5$ K, with a bolometric luminosity of $L_* \sim 10^{38}$ erg s^{-1}. For energies above 300 GeV, the highest absorption is produced with photons around 1.7 eV, i.e. in the NIR band, emitted by both companion and jet. Following Aharonian (2004), the cross-section of this interaction would be $\sigma_{\gamma\gamma} \sim \frac{2}{3}\sigma_{Thomson}\omega_0^{-1}ln(\omega_0) \sim 1 \times 10^{-25}$ cm^2, where $\sigma_{Thomson}$ is the cross-section of the IC process in the Thomson regime (see Sect. 1.2.1.5) and $\omega_0 = E_{\gamma 1} \cdot E_{\gamma 2}$ the product of the energies of the two photons colliding (300 GeV and 1.7 eV, in units of $m_e c^2$). This way, the absorption can be estimated as $\tau \sim \sigma_{\gamma\gamma} \cdot n_{NIR} \cdot r$ (Aharonian et al. 2005), where r is the size of the emitting region and $n_{NIR} \sim L_{NIR}/(4\pi r^2 c E_{\gamma 2})$, with c the speed of light and $E_{\gamma 2}$ the target photons energy. For the calculation of n_{NIR} I assume $L_{NIR} = L_* = 10^{38}$ erg s^{-1} (i.e. most of the NIR contribution comes from the donor star) and $E_{\gamma 2} = 2.7 \times 10^{-12}$ erg. Thus, it would not be until a radius of $\sim 10^{13}$ cm, i.e. outside of the binary scale ($R_{orb,CygX-3} \sim 2.5 \times 10^{11}$ cm) that the absorption does not affect VHE c. The independent *Fermi*-LAT analysis performed for the 2016 flare yielded a detection between 300 MeV and 20 GeV, approximately. To avoid absorption by UV and X-rays from the disk, HE gamma-ray photons in Cygnus X-3 had to be produced at distances above 10^{10}–10^{11} cm from the compact object, still inside the binary scales where the interaction with the stellar photon field gives rise to orbital modulated emission. Thus, given the MAGIC non-detection, VHE emission, if produced, could be expected inside the jets at a distance $<10^{13}$ cm, maybe related to the HE emission site. On the other hand, during the August–September campaign, MAGIC observed Cygnus X-3 during the highest radio flux ($\sim$9.5 Jy) on MJD 57651 for 4.8 h. This day corresponds to moon observations, in which the NSB reached the level of $12 - 18\times$ NSB$_{dark}$. No significant signal or emission was found during this day, which could reinforce the idea that VHE gamma rays are originated inside the binary scale and not at the radio-emitting regions of the jets far from the compact object. Note, however, that with the mentioned effective time, the sensitivity of the current MAGIC telescopes would be around 2.1% Crab Units (C.U.), to which an additional $\gtrsim$10% degradation has to be considered given the moonlight (MAGIC Collaboration et al. 2017). Therefore, it would be necessary to observe the source during longer periods simultaneously with radio outbursts to provide strong conclusions regarding the VHE production region.

In Fig. 5.5, the SED combining both HE and VHE results for the 2016 flare is shown. *Fermi*-LAT butterfly corresponds to the spectrum obtained by the independent analysis for *Period 2*, in which emission above $\sim$20 GeV is not detected. No HE cutoff is obtained in this *Fermi*-LAT analysis. There are two facts that are worth being highlighted: first of all, this SED allows us to probe the improvement achieved on the MAGIC performance after the upgrade of 2011–2012 (see Sect. 2.4). Former and current ULs were obtained assuming the same spectral shape and with very similar effective time ($\sim$60 h in Aleksić et al. 2010b and $\sim$70 h in this campaign). The main difference is that the pre-upgrade stand-alone MAGIC I telescope was used for observations shown in Aleksić et al. (2010b), while here results in post-upgrade stereoscopic mode were presented. Consequently, there is a large difference in sensitivities between both campaigns. While during mono observations, the sensitivity was roughly $\sim$1.6% C.U. above 270 GeV in 50 h, currently the sensitivity is $\sim$0.66% C.U. above 220 GeV for the same time. The latter is just marginally affected by moon observations, around $\gtrsim$10% worst sensitivity with nominal HV (MAGIC Collaboration et al. 2017), given that more than 50 h were observed in dark conditions. The sensitivity difference between both campaigns is remarkable at low energies, where the NSB affects the most our observations. Spurious triggered signals by the NSB are highly reduced in stereoscopic mode at the L3 trigger, which implies a better sensitivity as evidenced in Fig. 5.5. Moreover, results from Aleksić et al. (2010b) were obtained with the two oldest readout systems, *Siegen* and MUX. The former presented a very slow readout speed (300 MSamples/s) compared to that of MUX and the currently one used nowadays (2 GSamples/s and 1.64 GSamples/s, respectively). This slow integration time increased the integrated NSB, worsening the sensitivity up to 40% (see Sect. 2.4.1.5).

Secondly, the deep campaign performed during August–September 2016 with almost 70 h of good quality data allowed us to obtain very constraining ULs at 95% CL. Such low ULs are already at the level or below the 50 hours-sensitivity curve from CTA-North[1] at energies $<$1 TeV. Therefore, large observation time would be needed to detect Cygnus X-3 at low energies even with the next, more sensitive, generation of IACTs. This can be understood in two ways: one possibility is that the

---

[1]Taken from https://www.cta-observatory.org/science/cta-performance/.

VHE flux from Cygnus X-3 is extremely low given the unusually high absorption and consequently, great amount of time would be necessary with CTA under flaring activity to shed light on this regime. In this case, other microquasar with smaller opacity to gamma rays, as Cygnus X-1, are favored to be detected by CTA. Another possible scenario is the existence of a cutoff between HE (detected up to ~20 GeV) and VHE. The nature of this speculative cutoff could be either the absorption itself or a possible inefficiency inside the Cygnus X-3 jets. Nevertheless, the extrapolation of the *Fermi*-LAT spectrum, consistent with the ULs obtained in this thesis, seems to make unlikely the second scenario, although a more dedicated *Fermi*-LAT analysis could bring more information.

References

Ackermann M, Ajello M, Allafort A, Baldini L, Ballet J, Barbiellini G, Bastieri D, Belfiore, Bellazzini R, Berenji, Blandford RD, Bloom ED, Bonamente E, Borgland AW, Bottacini E, Brigida M, Bruel P, Buehler R, Buson S, Caliandro GA, Cameron RA, Caraveo PA, Casandjian JM, Cecchi C, Chekhtman A, Cheung CC, Chiang J, Ciprini S, Claus R, Cohen-Tanugi J, de Angelis A, de Palma F, Dermer CD, do Couto e Silva E, Drell PS, Dumora, Favuzzi C, Fegan, Focke, Fortin, Fukazawa, Fusco, Gargano F, Germani S, Giglietto N, Giordano F, Giroletti M, Glanzman T, Godfrey G, Grenier IA, Guillemot, Guiriec, Hadasch D, Hanabata Y, Harding AK, Hayashida M, Hayashi K, Hays E, Jóhannesson G, Johnson AS, Kamae T, Katagiri, Kataoka J, Kerr, Knödlseder J, Kuss M, Lande J, Latronico L, Lee, Longo F, Loparco F, Lott, Lovellette MN, Lubrano P, Martin, Mazziotta MN, McEnery JE, Mehault J, Michelson PF, Mitthumsiri W, Mizuno T, Monte, Monzani ME, Morselli A, Moskalenko IV, Murgia S, Naumann-Godo, Nolan, Norris, Nuss, Ohsugi T, Okumura, Orlando E, Ormes JF, Ozaki, Paneque D, Parent, Pesce-Rollins M, Pierbattista, Piron F, Pohl, Prokhorov, Rainò S, Rando R, Razzano M, Reposeur, Ritz S, Parkinson, Sgrò, Siskind EJ, Smith DA, Spinelli P, Strong AW, Takahashi T, Tanaka T, Thayer JG, Thayer JB, Thompson DJ, Tibaldo O, Torres, Tosti, Tramacere, Troja E, Uchiyama Y, Vandenbroucke J, Vasileiou V, Vianello G, Vitale V, Waite AP, Wang, Winer BL, Wood KS, Yang Z, Zimmer S, Bontemps Ackermann M et al. (2011) Science, 334, 1103

Aharonian FA (2004) Very high energy cosmic gamma radiation: a crucial window on the extreme Universe. World Scientific Publishing Co

Aharonian F, Akhperjanian AG, Aye KM, Bazer-Bachi AR, Beilicke M, Benbow W, Berge D, Berghaus P, Bernlöhr K, Boisson C, Bolz O, Borrel V, Braun L, Breitling F, Brown AM, Gordo J, Chadwick PM, Chounet LM, Cornils R, Costamante L, Degrange B, Dickinson HL, Djannati-Ataï A, Drury L, Dubus G Emmanoulopoulos D, Espigat P, Feinstein F, Fleury P, Fontaine G, Fuchs Y, Funk S, Gallant YA, Giebels S, Gillessen S, Glicenstein JF, Goret P, Hadjichristidis C, Hauser M, Heinzelmann G, Henri G, Hermann G, Hinton JA, Hofmann W, Holleran M, Horns D, Jacholkowska A, de Jager OC, Khélifi B, Komin NU, Konopelko A, Latham IJ, Le Gallou R, Lemière A, Lemoine-Gourmard M, Leroy N, Lohse T, Marcowith A, Martin JM, Martineau-Huynh O, Masterson C, McComb TJL, de Naurois M, Nolan SL, Noutsos A, Orford KJ, Osborne JL, Ouchrif M, Panter M, Pelletier G, Pita S, Pühlhofer G, Punch M, Raubenheimer BC, Raue M, Raux J, Rayner SM, Reimer A, Reimer O, Ripken J, Rob L, Rolland L, Rowell G, Sahakian V, Saugé L, Schlenker S, Schlickeiser R, Schuster C, Schwanke U, Siewert M, Sol H, Spangler D, Steenkamp R, Stegmann C, Tavernet JP, Terrier R, Théoret CG, Tluczykont M, Vasileiadis G, Venter C, Vincent P, Völk HJ, Wagner SJ, Aharonian F et al (2005) Science 309:746

Aleksić, Antonelli, Antoranz, Backes, Baixeras, Barrio, Bastieri, Becerra González, Bednarek, Berdyugin, Berger, Bernardini, Biland, Blanch, Bock, Boller, Bonnoli, Bordas, Borla Tridon, Bosch-Ramon, Bose, Braun, Bretz, Britzger, Camara, Carmona, Carosi, Colin, Contreras, Cortina,

Costado, Covino, Dazzi, De Angelis, De Cea del Pozo, De Lotto, De Maria, De Sabata, Delgado Mendez, Doert, Domínguez, Dominis Prester, Dorner, Doro, Elsaesser, Errando, Ferenc, Fonseca, Font, García López, Garczarczyk, Gaug, Godinovic, Göebel, Hadasch, Herrero, Hildebrand, Höhne-Mönch, Hose, Hrupec, Hsu, Jogler, Klepser, Krähenbühl, Kranich, La Barbera, Laille, Leonardo, Lindfors, Lombardi, Longo, López, Lorenz, Majumdar, Maneva, Mankuzhiyil, Mannheim, Maraschi, Mariotti, Martínez, Mazin, Meucci, Miranda, Mirzoyan, Miyamoto, Moldón, Moles, Moralejo, Nieto, Nilsson, Ninkovic, Orito, Oya, Paiano, Paoletti, Paredes, Partini, Pasanen, Pascoli, Pauss, Pegna, Perez-Torres, Persic, Peruzzo, Prada, Prandini, Puchades, Puljak, Reichardt, Rhode, Ribó, Rico, Rissi, Rügamer, Saggion, Saito, Saito, Salvati, Sánchez-Conde, Satalecka, Scalzotto, Scapin, Schultz, Schweizer, Shayduk, Shore, Sierpowska-Bartosik, Sillanpää, Sitarek, Sobczynska, Spanier, Spiro, Stamerra, Steinke, Struebig, Suric, Takalo, Tavecchio, Temnikov, Terzic, Tescaro, Teshima, Torres, Vankov, Wagner, Weitzel, Zabalza, Zandanel, Zanin, MAGIC Collaboration Aleksić J et al (2010b) 721, 843

Archambault M, Beilicke M, Benbow W, Berger K, Bird R, Bouvier A, Buckley JH, Bugaev V, Byrum K, Cerruti M, Chen X, Ciupik L, Connolly MP, Cui W, Duke C, Dumm J, Errando M, Falcone A, Federici S, Feng Q, Finley JP, Fortson L, Furniss A, Galante N, Gillanders GH, Griffin S, Griffiths ST, Grube J, Gyuk G, Hanna D, Holder J, Hughes G, Humensky TB, Kaaret P, Kertzman M, Khassen Y, Kieda D, Krawczynski H, Lang MJ, Madhavan AS, Maier G, Majumdar P, McArthur S, McCann A, Moriarty P, Mukherjee R, Nieto D, O'Faoláin de Bhróithe A, Ong RA, Otte AN, Pandel D, Park N, Perkins JS, Pohl M, Popkow A, Prokoph H, Quinn J, Ragan K, Rajotte J, Reyes LC, Reynolds PT, Richards GT, Roache E, Sembroski GH, Sheidaei, Smith, Staszak, Telezhinsky, Theiling, Tucci, Tyler, Varlotta, Vincent, Wakely, Weekes, Weinstein, Williams, Zitzer, VERITAS Collaboration and, McCollough, Astrophysical Observatory Archambault S et al (2013) 779, 150

Becklin E, Neugebauer G, Hawkins FJ, Mason KO, Sanford W, Matthews K, Wynn-Williams CG, Becklin EE et al (1973) 245, 302

Bhat CL, Sapru ML, Razdan H, Bhat CL et al (1986) 306, 587

Carpano S, Pollock AMT, Wilms J, Ehle M, Schirmer M, Carpano S et al (2007) 461, L9

Cheung CC et al (2016) The Astronomer's telegram, 9502

Collaboration Fermi LAT, Abdo AA, Ackermann M, Ajello M, Axelsson, Baldini L, Ballet J, Barbiellini G, Bastieri D, Baughman Bechtol K, Bellazzini R, Berenji B, Blandford RD, Bloom ED, Bonamente E, Borgland AW, Brez A, Brigida M, Bruel P, Burnett TH, Buson S, Caliandro GA, Cameron RA, Caraveo PA, Casandjian JM, Cecchi C, Çelik O, Chaty S, Cheung CC, Chiang J, Ciprini S, Claus R, Cohen-Tanugi J, Cominsky, Conrad Corbel S, Corbet R, Dermer CD, de Palma F, Digel SW, do Couto e Silva E, Drell PS, Dubois R, Dubus G, Dumora D, Farnier, Favuzzi C, Fegan SJ, Focke, Fortin P, Frailis M, Fusco P, Gargano F, Gehrels N, Germani S, Giavitto, Giebels, Giglietto N, Giordano, Glanzman T, Godfrey G, Grenier IA, Grondin MH, Grove JE, Guillemot, Guiriec S, Hanabata, Harding AK, Hayashida M, Hays E, Hill AB, Hjalmarsdotter, Horan D, Hughes RE, Jackson, Jóhannesson G, Johnson AS, Johnson RP, Johnson,TJ Kamae T, Katagiri H, Kawai, Kerr M, Knödlseder J, Kocian, Koerding E, Kuss M, Lande J, Latronico L, Lemoine-Goumard M, Longo F, Loparco F, Lott B, Lovellette MN, Lubrano P, Madejski, Makeev A, Marchand, Marelli, Max-Moerbeck, Mazziotta, McColl, McEnery JE, Meurer C, Michelson PF, Migliari, Mitthumsiri W, Mizuno T, Monte C, Monzani ME, Morselli A, Moskalenko IV, Murgia S, Nolan PL, Norris JP, Nuss E, Ohsugi T, Omodei N, Ong, Ormes JF, Paneque D, Parent D, Pelassa V, Pepe M, Pesce-Rollins M, Piron F, Pooley, Porter TA, Pottschmidt, Rainò S, Rando R, Ray PS, Razzano M, Rea N, Readhead, Reimer A, Reimer O, Richards, Rochester, Rodriguez, Rodriguez, Romani RW, Ryde F, Sadrozinski FW, Sander A, Saz Parkinson PM, Sgrò C, Siskind EJ, Smith A, Smith PD, Spinelli P, Starck JL, Stevenson, Strickman MS, Suson DJ, Takahashi H, Tanaka T, Thayer JG, Thompson, DJ, Tibaldo L, Tomsick, Torres DF, Tosti G, Tramacere A, Uchiyama Y, Usher TL, Vasileiou V, Vilchez N, Vitale V, Waite AP, Wang P, Wilms J, Winer, Wood, Ylinen, Ziegler Fermi LAT Collaboration et al (2009b) Science 326:1512

Corbel S, Dubus G, Tomsick J, Szostek A, Corbet R, Miller-Jones J, Richards JL, Pooley G, Trushkin S, Dubois R, Hill AB, Kerr M, Max-Moerbeck W, Readhead ACS, Bodaghee A, Tudose V, Parent D, Wilms J, Pottschmidt Corbel S et al (2012), 421, 2947

Egron E, Pellizzoni A, Giroletti M, Righini S, Orlati A, Iacolina N, Navarrini, Buttu M, Migoni C, Melis A, Concu R, Vargiu, Bachetti M, Pilia M, Trois A, Loru S, Marongiu M, Egron E et al (2016) The Astronomer's telegram, 9508

Giacconi R, Gorenstein P, Gursky H, Waters JR et al. (1967) 148: L119

Gregory PC, Kronberg PP, Seaquist ER, Hughes VA, Woodsworth A, Viner MR, Retallack D, Hjellming RM, Balick B, Gregory PC et al (1972) Nat Phys Sci 239, 114

Hanson MM, Still MD, Fender Hanson RP et al (2000) 541, 308

Hjalmarsdotter L, Zdziarski AA, Larsson S, Beckmann V, McCollough M, Hannikainen DC, Vilhu Hjalmarsdotter L et al (2008) 384, 278

Krimm HA, Holland ST, Corbet RHD, Pearlman AB, Romano P, Kennea JA, Bloom JS, Barthelmy SD, Baumgartner WH, Cummings JR, Gehrels N, Lien AY, Markwardt CB, Palmer DM, Sakamoto T, Stamatikos M, Ukwatta Krimm HA et al (2013) 209, 14

MAGIC Collaboration, Ahnen ML, Ansoldi S, Antonelli, LA Arcaro C, Babić A, Banerjee B, Bangale P, Barres de Almeida U, Barrio JA, Becerra González J, Bednarek W, Bernardini E, Berti A, Bhattacharyya, Biasuzzi B, Biland A, Blanch O, Bonnefoy S, Bonnoli G, Carosi R, Carosi A, Chatterjee A, Colin P, Colombo E, Contreras JL, Cortina J, Covino S, Cumani P, Da Vela P, Dazzi F, De Angelis A, De Lotto B, de Oña Wilhelmi E, Di Pierro F, Doert M, Domínguez A, Dominis Prester D, Dorner D, Doro M, Einecke S, Eisenacher Glawion D, Elsaesser D, Engelkemeier D, Fallah Ramazani V, Fernández-Barral A, Fidalgo D, Fonseca MV, Font L, Fruck C, Galindo D, García López RJ, Garczarczyk M, Gaug M, Giammaria P, Godinović N, Gora D, Griffiths S, Guberman D, Hadasch D, Hahn A, Hassan T, Hayashida M, Herrera J, Hose J, Hrupec D, Hughes G, Ishio K, Konno Y, Kubo H, Kushida J, Kuveždić D, Lelas D, Lindfors E, Lombardi S, Longo F, López M, Maggio, Majumdar P, Makariev M, Maneva G, Manganaro M, Mannheim K, Maraschi L, Mariotti M, Martínez M, Mazin D, Menzel U, Minev M, Mirzoyan R, Moralejo A, Moreno V, Moretti E, Neustroev V, Niedzwiecki A, Nievas Rosillo M, Nilsson K, Ninci, Nishijima K, Noda K, Nogués L, Paiano S, Palacio J, Paneque D, Paoletti R, Paredes JM, Paredes-Fortuny X, Pedaletti G, Peresano M, Perri L, Persic M, Prada Moroni PG, Prandini E, Puljak I, Garcia JR, Reichardt I, Rhode W, Ribó M, Rico J, Saito T, Satalecka K, Schroeder S, Schweizer T, Shore SN, Sillanpää A, Sitarek J, Šnidarić I, Sobczynska D, Stamerra A, Strzys M, Surić T, Takalo L, Tavecchio F, Temnikov P, Terzić T, Tescaro D, Teshima M, Torres DF, Torres-Albà N, Treves A, Vanzo G, Vazquez Acosta M, Vovk I, Ward JE, Will M, Zarić MAGIC Collaboration et al (2017) ArXive-prints

Martí J, Paredes JM, Peracaula M, Martí J et al (2001) 375:476

Matsuoka M, Kawasaki K, Ueno S, Tomida H, Kohama M, Suzuki M, Adachi Y, Ishikawa M, Mihara T, Sugizaki M, Isobe N, Nakagawa Y, Tsunemi H, Miyata E, Kawai N, Kataoka J, Morii M, Yoshida A, Negoro H, Nakajima M, Ueda Y, Chujo H, Yamaoka K, Yamazaki O, Nakahira S, You T, Ishiwata R, Miyoshi S, Eguchi S, Hiroi K, Katayama H, Ebisawa K, Matsuoka M et al (2009) 61, 999

McCollough ML, Corrales L, Dunham MM, McCollough ML et al (2016) 830, L36

Miller-Jones JCA, Blundell KM, Rupen MP, Mioduszewski AJ, Duffy P, Beasley AJ, Miller-Jones JCA et al (2004) 600, 368

Parsignault DR, Gursky H, Kellogg EM, Matilsky T, Murray S, Schreier S, Tananbaum H, Giacconi R, Brinkman AC, Parsignault DR et al (1972) Nat Phys Sci 239, 123

Piano C, Tavani M, Bulgarelli A, Verrecchia F, Donnarumma I, Munar-Adrover, Minervini, Fioretti, Zoli, Pittori, Lucarelli, Vercellone, Striani, Cardillo, Gianotti, Trifoglio, Giuliani, Mereghetti, Caraveo, Perotti, Chen, Argan, Costa, Del Monte, Evangelista Y, Feroci, Lazzarotto F, Lapshov I, Pacciani L, Soffitta P, Sabatini S, Vittorini V, Pucella G, Rapisarda M, Di Cocco G, Fuschino F, Galli M, Labanti C, Marisaldi M, Pellizzoni A, Pilia M, Trois A, Barbiellini G, Vallazza E, Longo F, Morselli A, Picozza P, Prest M, Lipari P, Zanello D, Cattaneo PW, Rappoldi A, Colafrancesco

S, Parmiggiani N, Ferrari A, Antonelli A, Giommi P, Salotti L, Valentini G, D'Amico Piano G et al (2016) The Astronomer's telegram, 9429

Prestwich AH, Kilgard R, Crowther PA, Carpano S, Pollock AMT, Zezas A, Saar SH, Roberts TP, Ward MJ, Prestwich AH et al (2007) 669, L21

Singh, NS, Naik S, Paul B, Agrawal PC, Rao AR, Singh A, Singh NS et al (2002) 392, 161

Stark MJ et al (2003) 587, L101

Szostek A, Zdziarski AA, McCollough ML Szostek A et al (2008) 388, 1001

Tavani M, Bulgarelli G, Piano G, Sabatini S, Striani E, Evangelista Y, Trois A, Pooley G, Trushkin S, Nizhelskij NA, McCollough M, Koljonen KII, Pucella G, Giuliani A, Chen AW, Costa E, Vittorini V, Trifoglio M, Gianotti F, Argan A, Barbiellini G, Caraveo P, Cattaneo PW, Cocco V, Contessi T, D'Ammando F, Del Monte E, de Paris G, Di Cocco G, di Persio G, Donnarumma I, Feroci M, Ferrari A, Fuschino F, Galli M, Labanti C, Lapshov I, Lazzarotto F, Lipari P, Longo E, Mattaini E, Marisaldi M, Mastropietro M, Mauri A, Mereghetti S, Morelli E, Morselli A, Pacciani L, Pellizzoni A, Perotti F, Picozza P, Pilia M, Prest M, Rapisarda M, Rappoldi A, Rossi E, Rubini A, Scalise E, Soffitta P, Vallazza E, Vercellone S, Zambra A, Zanello D, Pittori, Verrecchia F, Giommi P, Colafrancesco S, Santolamazza P, Antonelli A, Salotti Tavani M et al (2009b) 462:620

Trushkin SA, Nizhelskij NA, Tsybulev PG, Zhekanis GV, Trushkin SA et al (2016a) The Astronomer's telegram, 9416

Trushkin SA, Nizhelskij NA, Tsybulev PG, Zhekanis GV, Trushkin SA et al (2016b) ArXiv e-prints

Trushkin SA, Nizhelskij NA, Tsybulev PG, Zhekanis GV, Trushkin SA et al (2016c) The Astronomer's telegram, 9501

van Kerkwijk MH, Charles PA, Geballe TR, King DL, Miley GK, Molnar LA, van den Heuvel EPJ, van der Klis M, van Paradijs J, van Kerkwijk MH et al (1992) 355, 703

Vladimirsky BM, Stepanian AA, Fomin VP, Vladimirsky BM et al (1973) International Cosmic Ray Conference, 1, 456

Waltman EB, Fiedler RL, Johnston KJ, Ghigo Waltman EB et al (1994) 08, 179

Waltman EB, Ghigo FD, Johnston KJ, Foster RS, Fiedler JH, Spencer Waltman EB et al (1995) 110, 290

Waltman SJ, Foster RS, Pooley GG, Fender RP, Ghigo FD, Waltman EB et al (1996) 112, 2690

Chapter 6
V404 Cygni

6.1 History

V404 Cygni is a low-mass microquasar composed of a 8–15 $M_\odot$ BH (Shahbaz et al. 1994) and a $0.7^{+0.3}_{-0.2} M_\odot$ K3 III companion star (Casares and Charles 1994; Khargharia et al. 2010). The system is located at a distance of 2.39 ± 0.14 kpc (Miller-Jones et al. 2009) in the Cygnus Constellation. It displays an orbital period of 6.5 days (Casares et al. 1992) and a jet inclination angle with respect to our line of sight of $\sim 67^{\circ+3}_{-1}$ (Khargharia et al. 2010).

There are reports from the 18th century in which V404 Cygni was considered by astronomers as a variable star. It was firmly detected in the optical in 1938, and afterwards in 1956, although misclassified as a nova event. It would not be until 1989, when the source entered in an outburst activity period (releasing great amount of energy in radio, optical and X-rays, as seen by the *Ginga* satellite, Makino et al. 1989), when it was finally classified as a LMXB. This way, the source seems to undergo extreme outbursts every two or three decades. Since 1989, it had remained in a quiescence state until June 2015.

6.1.1 June 2015 Outburst

On the 15th of June 2015, the hard X-ray satellite *Swift*-BAT detected a huge outburst from the direction of V404 Cygni, which sent a worldwide alert via the GCN (Barthelmy et al. 2015). This burst-like activity period was similar to that on 1989 but shorter, with a duration of ~ 11 days. During these days, the microquasar displayed multiple flares in a time-scale of hours and it became the brightest X-ray source in the sky, reaching fluxes larger than 30 C.U. between 20–40 keV (see Fig. 6.1). This exceptional behaviour led to a multiwavelength observational campaign from radio (Mooley et al. 2015), through hard X-rays with *INTEGRAL*

© Springer Nature Switzerland AG 2018

A. Fernández Barral, *Extreme Particle Acceleration in Microquasar Jets and Pulsar Wind Nebulae with the MAGIC Telescopes*, Springer Theses, https://doi.org/10.1007/978-3-319-97538-2_6

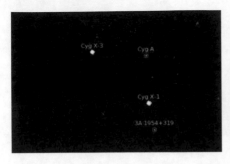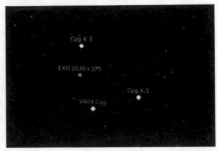

Fig. 6.1 V404 Cygni FoV as seen by *INTEGRAL*-IBIS (15 keV–10 MeV) before (*left*) and after (*right*) the outburst on the 19th of May 2015. The position of the source is marked with a cross and labeled in the *left* and *right* panels, respectively. Credit: C. Ferrigno, *INTEGRAL* Science Data Center

(Ferrigno et al. 2015) and HE gamma rays with *Fermi*-GBM (Younes 2015), up to VHE, observed by VERITAS (Archer et al. 2016) and MAGIC.

Specially interesting were the results obtained from the observations performed by *INTEGRAL* (Siegert et al. 2016) between the 17th and the 30th of June. Data from *INTEGRAL*-SPI were analyzed in three periods of three days each one (during flaring activity), in which an excess around 511 keV was detected. This excess was compatible with electron/positron annihilation. Siegert et al. (2016) showed that for energies below 200 keV, the hard X-ray spectrum could be described with the common thermal Comptonization, however above this energy an excess stood out. By adding a model spectrum of $e^{\pm}$ plasma with temperature T to the Comptonization model, the overall fit improved 5σ. The value of T (which describes the width of the annihilation line) differed considerably in each epoch, from $T \sim 1$ to ~ 200 keV, as seen in Fig. 6.2. The annihilation line emission is expected close by luminous accreting BH when the spectra of the $e^{\pm}$ extend above 511 keV, produced by the interaction between MeV gamma rays. This process becomes efficient given the small size of the V404 Cygni system. This observed annihilation line corresponded to a positron production rate of $\sim 10^{42}$ s^{-1}. Such a positron production rate found in V404 Cygni could support two interesting theories: first of all, that the microquasars could be the main producers of the $e^{\pm}$ plasma responsible for the diffuse annihilation radiation in the bulge region of the Galaxy and, secondly, that they could also be the main origin of the observed MeV continuum excess present in the inner parts of our Galaxy.

On the other hand, Loh et al. (2016) found an excess of 4σ using *Fermi*-LAT data, coincident with a giant radio flare (Trushkin et al. 2015a). Given the simultaneity of the 511 keV excess and the hint in the HE gamma-ray regime, the production site of latter is expected to be outside of the corona, most likely in the jets: the plasma is continuously emitting annihilation radiation (given rise to broad annihilation lines, see Fig. 6.2) increasing the opacity for photons with energies above 100 MeV which would be absorbed by the X-rays in the coronal region.

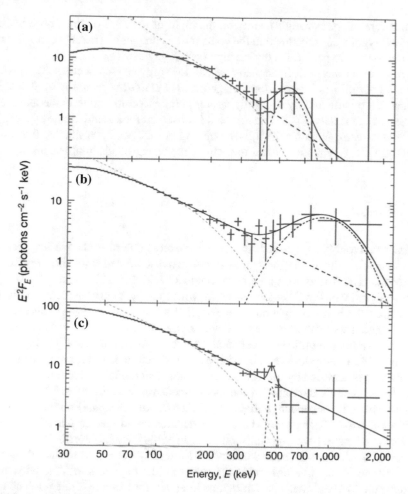

Fig. 6.2 Spectra of V404 Cygni in the soft gamma-ray regime obtained with *INTEGRAL* in three flaring periods between 17th and 30th of June 2015. Blue points (with 1σ error bars) correspond to the data, fitted to the Comptonization model (black dashed curve) and annihilation radiation (red dashed curve). Credit: Siegert et al. (2016)

6.2 Observations and Data Analysis

MAGIC observed V404 Cygni for 8 non-consecutive nights under dark conditions ($DC_{MI} < 2\,\mu A$) from the 18th to the 27th of June 2015, reaching a total of $\sim$10 h. Most of our observations were based on *INTEGRAL* alerts arrived from the GCN. The first alert, received on the 18th at 00:08:39 UT, presented such a high level flux that the source was believed to be a GRB. Therefore, the MAGIC telescopes pointed automatically to the coordinates of the source, following the GRB procedure (explained in Sect. 2.4.1.6). This scenario would be repeated along the entire period (with several alerts per night). Only on the night of the 23rd of June, observations of V404 Cygni were scheduled in the context of a campaign. Therefore, most of our observations were performed during high hard X-ray activity.

The analysis was performed using the MARS software (see Sect. 2.4.3). The first subruns (∼2 min) of observations triggered by *INTEGRAL* alerts were discarded as they can be affected by a slightly wrong pointing position due to the automatic movement of the telescopes. The source was observed for a zenith range of 5–50° in the *wobble-mode* (Fomin et al. 1994), pointing at four positions situated 0.4° away from the V404 Cygni coordinates to evaluate the background simultaneously.

6.3 Results

In order to avoid being spoiled by a putative excess due to trials, we searched for time intervals with the highest hard X-ray activity, related with the presence of relativistic jets. To do so, publicly available *INTEGRAL*-IBIS data (20–40 keV) were analyzed with the OSA software,[1] for which the light curve depicted in Fig. 6.3 was obtained. Onto this light curve, we applied a Bayesian block analysis (Scargle et al. 2013). This method is used to detect signal structures over the time with the aim of separating significant local[2] features from random errors. By applying the least possible assumptions (e.g. the shape of the signal) to avoid limiting the model, the division is achieved by fitting a piecewise constant model (a step function) to the data. Consequently, the range of the independent variable (time) will be split into bins in which the dependent variable (the intensity of the signal) remains constant within the errors. In our case, these local features could lead to periods in which particle acceleration was specially efficient in V404 Cygni. Nevertheless, none of the *INTEGRAL* flares presented exceptional intensity compared to the others (see Fig. 6.3) and hence, five time intervals, covering the highest flaring activity, were used for the MAGIC analysis. These time intervals are listed in Table 6.1. The total MAGIC data covered in these 5 periods defined with the Bayesian Block algorithm reach ∼7 h. However, we found no detection for this data sample, which led to a significance of 0.08σ.

For completeness, we searched for steady emission in the whole data sample and, given the high variability shown by the source, we also looked for emission in a daily basis analysis. None of these subsample yielded to a significant excess (see Table 6.2). We therefore computed integral and differential ULs using the Rolke et al. (2005) method, assuming a power-law spectrum of photon index $\Gamma = 2.6$ and 30% systematic uncertainty in the effective area of the gamma rays. The results, given for a 95% C.L., are quoted in Table 6.2. Differential ULs for the data within the Bayesian Blocks periods are shown in Fig. 6.4.

---

[1] http://www.isdc.unige.ch/integral/analysis.

[2] The local features are those located in a sub-range of time which do not repeat continuously, contrary to the *global* ones, as e.g. the periodicity.

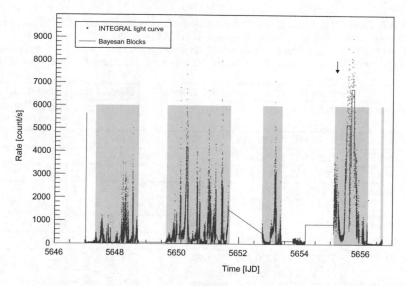

Fig. 6.3 Light curve of *INTEGRAL*-IBIS data (red points, 20–40 keV) onto which the Bayesian Blocks method (Scargle et al. 2013) was applied (blue lines). The intervals with the highest flaring activity used in the analysis of MAGIC data are highlighted by grey bands. The time is given in *INTEGRAL* Julian Date, defined as IJD=MJD-51544.0. The black arrow corresponds to the observation time at which the 4σ hint with *Fermi*-LAT data was obtained (Loh et al. 2016). Credit: Ahnen et al. (2017)

Table 6.1 Time intervals used in the MAGIC analysis based on the Bayesian Block algorithm applied on *INTEGRAL* data. *Credit* Ahnen et al. (2017)

Start (MJD)	Stop (MJD)
57191.337	57192.725
57193.665	57195.700
57196.765	57197.389
57199.116	57200.212
57200.628	57200.695

The hint at the level of 4σ seen in the HE regime by Loh et al. (2016) with the analysis of *Fermi*-LAT data, temporally coincident with the brightest radio flare, presented a peak on the 26th of June, at MJD 57199.2 ± 0.1 (black arrow in Fig. 6.3). MAGIC observed V404 Cygni for around 1 h simultaneously to this *Fermi*-LAT excess (during observations between 57199.158–57199.204, see Table 6.2). Nevertheless, the extrapolation of the *Fermi*-LAT spectrum, given between 100 MeV and 100 GeV and described by a power-law function of index 3.5, is two orders of magnitude lower than our ULs computed with the same photon index (see Fig. 6.4).

Table 6.2 *From left to right*: MAGIC observation period in calendar and MJD, effective time, significance and integral flux ULs for energies between 200 and 1250 GeV, assuming a power-law function of index $\Gamma = 2.6$. The last two rows report the results for the whole data sample and for the sample obtained based on the Bayesian Blocks algorithm (see Sect. 6.3). *Credit* Ahnen et al. (2017)

Observation Period		Eff. Time	Significance	Integral UL
[yyyy mm dd]	[MJD]	[hr]	[σ]	[$\times 10^{-11}$photons cm^{-2}s^{-1}]
2015 06 18	57191.006–57191.146	2.99	−0.43	0.51
2015 06 19	57191.960–57192.055	1.9	−0.6	1.0
2015 06 21	57193.997–57194.025	0.66	1.57	4.4
2015 06 22	57195.021–57195.049	1.33	0.09	1.7
	57195.103–57195.134			
2015 06 23	57196.003–57196.124	2.74	−0.45	0.37
2015 06 26	57199.158–57199.204	1.03	−1.41	0.66
2015 06 27	57200.085–57200.115	1.97	−0.57	1.2
	57200.144–57200.202			
Full data sample	–	10.65	−0.88	0.22
Bayesian Block selection	–	6.88	−0.42	0.48

6.4 Discussion and Conclusions

MAGIC observed the low-mass microquasar V404 Cygni during a major burst-like period, the last one since 25 years ago. These observations represented a good opportunity to study accretion-ejection processes given the proximity of the BH and the well-determined parameters of the binary (as its distance and the masses of the compact object and donor star). Mostly triggered by hard X-ray alerts from *INTEGRAL*, we obtained a total of $\sim$10 h of good quality data. By applying a Bayesian Block model to search for local variability, we reduced this sample to $\sim$7 h. However, nor steady emission (at any subsample) or variable emission in a night-wise basis was found.

Given the integral UL (for energies between 200–1250 GeV) obtained for the Bayesian Block selection, 4.8×10^{-12} cm^{-2}s^{-1}, and a distance of 2.4 kpc (Miller-Jones et al. 2009), the gamma-ray luminosity UL is $\sim$2 $\times$ 10^{33}erg s^{-1}, much lower than the extreme energy flux emitted in the hard X-ray band (20–400 keV) of $\sim$2 $\times$ 10^{38} erg s^{-1} ($\sim$20% Ledd of a 9 M$_\odot$ BH, (Rodriguez et al. 2015)). This luminosity UL is, in turn, two orders of magnitude lower than the one obtained in the HE gamma-ray regime by Loh et al. (2016). On the other side, Tanaka et al. (2016) developed a model for the jet emission in V404 Cygni, following a blazar approach (assuming one-zone synchrotron plus SSC model), in which the total power carried by the relativistic outflows reach 7.0 $\times$ 10^{37} erg s^{-1}. Therefore, the conversion efficiency of jet power to VHE gamma-ray luminosity in this low-mass microquasar is below 0.003%.

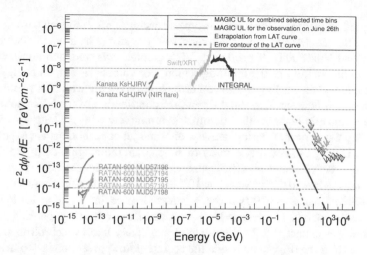

Fig. 6.4 Multiwavelength SED of V404 Cygni during the June 2015 flaring period. In red, MAGIC ULs are given for the combined Bayesian block time bins (7 h) for which a power-law function with photon index 2.6 was assumed. In green, MAGIC ULs for observations on June 26th, simultaneously taken with the *Fermi*-LAT hint (Loh et al. 2016). In this case, a photon index of 3.5 was applied following *Fermi*-LAT results. The extrapolation of the *Fermi*-LAT spectrum is shown in blue with 1σ contour (gray dashed lines). In the X-ray regime, *INTEGRAL* (20–40 keV, Rodriguez et al. (2015)) and *Swift*-XRT (0.2–10 keV, Tanaka et al. (2016)) data are depicted. At lower energies, Kanata-HONIR optical and NIR data are shown, taken from Tanaka et al. (2016). Finally, RATAN-600 radio data, from Trushkin et al. (2015b), are presented for different days along to the flaring activity period. Credit: Ahnen et al. (2017)

I compare V404 Cygni luminosity UL with the one obtained for the high-mass microquasar Cygnus X-1 (Sect. 4.3), given certain similarities shown by both microquasars. On one hand, both sources are located at similar distances (2.4 kpc in the case of V404 Cygni and 1.86 kpc for Cygnus X-1, Reid et al. (2011)). Secondly, the total power emitted by their relativistic outflows are of the same order (for V404 Cygni, the power jet reaches 7.0×10^{37} erg s^{-1}, as mentioned above, taken from Tanaka et al. (2016), whilst for Cygnus X-1 this is expected to be 10^{36}–10^{37} erg s^{-1}, (Gallo et al. 2005)). Finally, during the flare that gave rise to the hint in the HE gamma-ray band (reported by Loh et al. (2016)), V404 Cygni appeared to stay in a HS or IS, states during which GeV emission was detected in Cygnus X-1 (see Sect. 4.2). Despite all the resemblances, the conversion efficiency into gamma-ray luminosity from the jet power is one order of magnitude smaller in the case of V404 Cygni (0.003% compared to the 0.06% obtained for Cygnus X-1).

The fact that the only two detected microquasars in the gamma-ray regime, Cygnus X-1 (Sect. 4.2) and Cygnus X-3 (Tavani et al. 2009; Fermi LAT Collaboration et al. 2009), host hot and massive companion stars suggests that the most efficient mechanism to emit in this energy band is the IC on stellar photons. Low-mass microquasars are not suitable to fit in this scenario because of the cold and old companions (see Sect. 3.2.4.3), but they cannot be discarded as VHE emitters.

Models predict TeV emission from this type of systems under efficient particle acceleration on the jets (Atoyan and Aharonian 1999) or strong hadronic jet component (Romero and Vila 2008). If produced, VHE gamma rays may annihilate via pair creation in the vicinity of the emitting region. For gamma rays in an energy range between 200 GeV–1.25 TeV, the largest cross-section occurs with NIR photons. For a low-mass microquasar, like V404 Cygni, the contribution of the NIR photon field from the companion star (with a bolometric luminosity of $\sim 10^{32}$ erg s^{-1}) is very low. During the period of flaring activity, disk and jet contributions are expected to dominate. During the outburst activity of June 2015, the magnitude of the K-band reached m = 10.4 (Shaw et al. 2015), leading to a luminosity on the NIR regime of $L_{NIR} = \nu \phi_{m=0} 4\pi d^2 10^{-m/2.5} = 4.1 \times 10^{34}$ erg s^{-1}, where ν is the frequency for the 2.2 μm K-band, $\phi_{m=0} = 670$ Jy is the K-band reference flux and $d = 2.4$ kpc is the distance to the source. The detected NIR radiation from V404 Cygni during this flaring period, was expected to be dominated by optically-thick synchrotron emission from the jet or to be originated inside the accretion flow, given the lack of evidence of polarization (Tanaka et al. 2016). Consequently, stronger gamma-ray absorption is expected at the base of the jets. The gamma-ray opacity due to NIR radiation inside V404 Cygni can be estimated as $\tau_{\gamma\gamma} \sim \sigma_{\gamma\gamma} \cdot n_{NIR} \cdot r$, given by Aharonian et al. (2005). The cross-section of the interaction is defined by $\sigma_{\gamma\gamma}$, whose value is $\sim 10^{-25}$ cm^2. The NIR photon density is calculated as $n_{NIR} = L_{NIR}/\pi r^2 c\epsilon$, where r is the radius of the jet where NIR photons are expected to be emitted; c is the speed of light and $\epsilon \sim 1 \times 10^{-12}$ erg is the energy of the target photon field. Assuming the aforementioned luminosity of $L_{NIR} = 4.1 \times 10^{34}$ erg s^{-1}, the gamma-ray opacity at a typical radius $r \sim 1 \times 10^{10}$ cm may be relevant enough to avoid VHE emission above 200 GeV. Moreover, if IC on X-rays at the base of the jets ($r \lesssim 1 \times 10^{10}$ cm) is produced, this could already prevent electrons to reach the TeV regime, unless the particle acceleration rate in V404 Cygni is close to the maximum achievable including specific magnetic field conditions (see e.g. Khangulyan et al. (2008)). On the other hand, absorption becomes negligible for $r > 1 \times 10^{10}$ cm. Thus, if the VHE emission is produced in the same region as HE radiation ($r \gtrsim 1 \times 10^{11}$ cm, to avoid X-ray absorption), then it would not be significantly affected by pair production attenuation ($\sigma_{\gamma\gamma} < 1$). Therefore a VHE emitter at $r \gtrsim 1 \times 10^{10}$ cm, along with the non-detection by MAGIC, suggests either inefficient particle acceleration inside the V404 Cygni jets or not enough energetics of the VHE emitter.

Content included in this chapter has been published in Ahnen et al. (2017) (Monthly Notices of the Royal Astronomical Society ©: 2017. Published by Oxford University Press on behalf of the Royal Astronomical Society. All rights reserved).

References

Aharonian F et al (2005) Science 309:746
Ahnen ML et al (2017) 471, 1688
Archer A et al (2016) 831, 113

Atoyan AM et al (1999) 302, 253
Barthelmy SD et al (2015) GRB Coordinates Network, 17929
Casares J et al (1992) 355, 614
Casares J et al (1994) 271, L5
Collaboration Fermi LAT et al (2009) Science 326:1512
Ferrigno C et al (2015) The Astronomer's Telegram, 7662
Fomin VP et al (1994) Astropart Phys 2:137
Gallo E et al (2005) 436, 819
Khangulyan D et al (2008) 383, 467
Khargharia J et al (2010) 716, 1105
Loh A et al (2016) 462, L111
Makino F et al (1989) 4786
Miller-Jones JCA et al (2009) 394, 1440
Mooley K et al (2015) The Astronomer's Telegram, 7658
Reid MJ et al (2011) 742, 83
Rodriguez J et al (2015) 581, L9
Rolke WA et al (2005) Nucl Instrum Methods Phys Res A 551:493
Romero GE et al (2008) 485, 623
Scargle JD et al (2013) 764, 167
Shahbaz T et al (1994) 271, L10
Shaw AW et al (2015) The Astronomer's Telegram, 7738
Siegert T et al (2016) 531, 341
Tanaka YT et al (2016) 823, 35
Tavani M et al (2009) 462, 620
Trushkin SA et al (2015a) The Astronomer's Telegram, 7716
Trushkin SA et al (2015b) The Astronomer's Telegram, 7667
Younes G (2015) GRB Coordinates Network, 17932

Part III
Pulsar Wind Nebulae in the Very High-Energy Gamma-Ray Regime

Fig. III.1 Superimposed image of the X-ray emission from the Crab Pulsar (white) with the visible Crab Nebular (red). Credit: (Crab Pulsar image) NASA/CXC/SAO/F. D. Seward, W. H. Tucker and R. A. Fesen; (Crab Nebula image) Adam Block/Mount Lemmon SkyCenter/University of Arizona

Chapter 7
Introduction to Pulsar Wind Nebulae

The ejected material by a SN explosion interacts with the surrounding environment, giving rise to a new astrophysical object known as SNR (for a review of this matter, I refer the reader to (Reynolds 2008) and to Appendix A for more information on the first evolutionary phase). In turn, the possible leftovers product of this SN are very different depending on its initial mass, as described in Sect. 3.1.

SNRs can be classified into three types: *shell-like* **SNRs**, whose name arises from their shell-like structure (as for example, Cassiopeia A); **PWNe** or *plerions*, if the central object is a pulsar that constantly powers and injects particles into the remnant (as in the case of the famous Crab Nebula); and finally, *composite* **remnant**, produced when the PWN is surrounded in turn by a *shell-like* SNR (e.g., HESS J1818–154). In this chapter, I will focus on the components and expected emission from the second type, the PWNe.

7.1 Central Object in PWNe

7.1.1 Neutron Star

NSs are the most dense objects known in our Universe (Fig. 7.1). Born from the explosion of massive stars, these objects normally achieve diameters of ∼20 km and masses around 1.4–3 $M_\odot$, which leads to densities of $\rho \sim 10^{17}$ kg/m^3. Two properties are specially relevant in these objects: the angular momentum and the magnetic flux, both of them conserved from the initial massive star. As the massive star collapses into a much smaller object, the rotation rate needs to increase as a result of conservation of angular momentum. Therefore, NSs can experiment rotation rates or periods from 1 ms to 10 s. On the other hand, as the star's surface is smaller, the magnetic field will

© Springer Nature Switzerland AG 2018
A. Fernández Barral, *Extreme Particle Acceleration in Microquasar Jets and Pulsar Wind Nebulae with the MAGIC Telescopes*, Springer Theses, https://doi.org/10.1007/978-3-319-97538-2_7

Fig. 7.1 Artist's view of a
NS. Credit: Casey
Reed/Penn State University

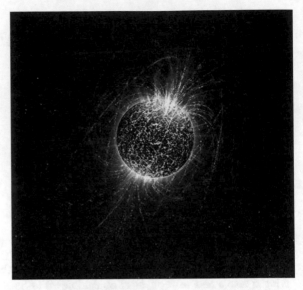

Fig. 7.2 Internal structure of
a NS. Credit: NASA/GSFC

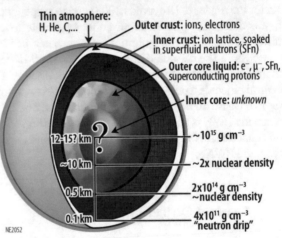

be stronger due to the magnetic flux conservation. As a result, a highly magnetized
rapidly-spinning NS is born. NSs are usually modeled as magnetic dipoles.

NSs are mainly composed of neutrons, although their internal structure is more
complicated (Fig. 7.2). Below a very thin atmosphere, NSs present an outer crust of
~300 m deep formed by ions and electrons. Deeper in the star, an inner crust of
~600 m deep is developed, which mixes neutrons and electrons. The bulk of the NS
remains in the so-called outer core, a ~9 km deep layer, composed of a superfluid
neutrons, as well as an small number of superconducting protons. The composition
of the innermost layer, the core of the NS, is still unclear, but some theories suggest
exotic solid matter consisting on elementary particles, such as quarks and gluons.

There are different types of NSs according to the primary source of their emission. In this thesis, we are interested in a sub-class known as *Rotation-Oowered Pulsars (RPPs)* or just pulsars, in which the emission arises from the rotational energy of the NS. For an extended review on NSs, see Harding (2013).

7.1.2 Pulsars

Pulsars are a type of NS that emit beams of electromagnetic radiation powered by rotational energy. This emission is detected when crossing our line of sight, producing this way the pulsed appearance (the so-called lighthouse effect). Their existence was postulated by Pacini (1967) to explain the Crab Nebula system, one year before their discovery in the radio band by Jocelyn Bell (Hewish et al. 1968).

Given that the primary source of radiation in these systems is their rotational energy, E_{rot}, in the following we will see the relation of their principal properties with the energy loss. For a comprehensive review on PWNe, the reader is referred to Gaensler and Slane (2006).

The main feature of pulsars is the so-called spin-down power, $\dot{E} = -dE_{rot}/dt$, i.e. the rate at which the rotational energy is released. It is defined as:

$$\dot{E} = 4\pi^2 I \frac{\dot{P}}{P^3} = I\omega\dot{\omega} \tag{7.1}$$

where P is the rotational or spin period and $\omega = 2\pi/P$ is the angular frequency of the NS ($\dot{P}$ and $\dot{\omega}$ correspond with their derivative with respect to time, respectively), and $I \sim 10^{38}$ kg m^2 is the moment of inertia of the star. This leads to $\dot{E}$ values for known pulsars ranging between 10^{28}–10^{38} erg/s.

The rotational period decreases with time with respect to its initial spin period P_0. This spin down can be define as $\dot{\omega} = -\kappa\omega^n$, where κ is a constant and n is the so-called *braking index*. The value of n has only been firmly measured for four pulsars, ranging between 1.4 and 2.9. It is usually considered $n = 3$ for the rest of pulsars, which corresponds to the spin down through magnetic dipole radiation. If we take this spin down as a function of the period, following (Manchester and Taylor 1977), then we get an expression for the age of the pulsar, τ:

$$\tau = \frac{P}{(n-1)\dot{P}} \left[1 - \left(\frac{P_0}{P}\right)^{n-1} \right] \tag{7.2}$$

Equation 7.2, for the case of $n = 3$ and $P_0 << P$, can be simplified as:

$$\tau_c = \frac{P}{2\dot{P}} \tag{7.3}$$

This τ_c is the so-called characteristic age of the pulsar. Normally, the characteristic age obtained through Eq. 7.3 overestimates the real age of the pulsar.

On the other hand, by following the magnetic dipole assumption (Ostriker and Gunn 1969), one can obtain the magnetic field on the surface of the NS, B in G, as a function of the period (in seconds):

$$B \sim \mu R \simeq 3.2 \times 10^{19} (P \dot{P})^{1/2} \tag{7.4}$$

where μ is the magnetic dipole moment and R is the radius of the NS. Common values for the surface magnetic field range between 10^{12}–10^{13} G.

The spin-down power and spin period will evolve with time. Assuming that the *braking index*, n, is constant, we can define the evolution as:

$$\dot{E}(t) = \dot{E}_0 \left(1 + \frac{t}{\tau_0} \right)^{-\frac{n+1}{n-1}} \tag{7.5}$$

$$P(t) = P_0 \left(1 + \frac{t}{\tau_0} \right)^{\frac{1}{n-1}} \tag{7.6}$$

where $\dot{E}_0$ corresponds to the initial value of the spin-down power and $\tau_0 = \frac{2\tau_c}{n-1} - t$ is the spin-down time scale.

7.1.2.1 Magnetosphere and Wind Zone

The charge (due to ions and electrons) in the surface of the NS induce an electric field parallel to the magnetic field, i.e. perpendicular to the surface. This electric field is so strong that pulls charges out of the surface against the gravitational force. These particles follow the magnetic field lines populating the surrounding of the NS, the so-called NS magnetosphere. In this magnetosphere, the charges co-rotate with the NS and their charge density is approximately the Goldreich-Julian one, $\rho_{GJ} \sim \Omega \cdot B/2\pi c$ (Goldreich and Julian 1969), where Ω is the angular speed and c is the speed of light. The NS magnetosphere is delimited by the so-called *light cylinder* (see Fig. 7.3), whose radius is the distance at which the co-rotating plasma and magnetic field speed equals the speed of light ($R_{LC} = c/\Omega$). Therefore, the open magnetic field lines, through which the charge particles leave the *light cylinder*, do not co-rotate with the NS. This flowing of particles outside of the magnetosphere creates zones with reduced charge density and hence, contrary to what happens under Goldreich-Julian charge density, an electric field (parallel to the magnetic field) can develop. This way, particles following the open magnetic field lines can be accelerated in certain regions and, due to the strong magnetic field, they will be able to emit synchrotron radiation. This radiation will in turn produce pairs of $e^{\pm}$, giving rise to electromagnetic cascades. However, these secondary pairs short out the electric field and consequently, constrain their own acceleration. This pulsar-

Fig. 7.3 Diagram of the canonical magnetic dipole model applied to pulsars. The magnetosphere of the NS is delimited by the light cylinder. Credit: Aliu et al. (2008)

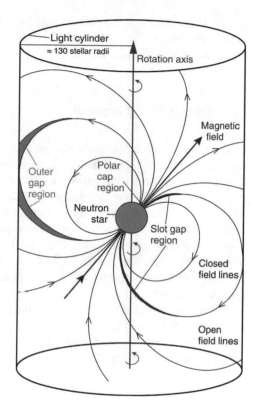

driven wind, during its expansion outwards from the light cylinder, decelerates and the pressure and interaction produced by ISM give rise to the so-called termination shock, in which $e^{\pm}$ get accelerated up to VHE. The region between the *light cylinder* and the termination shock is called *wind zone*.

The magnetization of the pulsar wind is given by σ:

$$\sigma \equiv \frac{F_{E \times B}}{F_{particles}} = \frac{B^2}{4\pi \rho \gamma c^2} \qquad (7.7)$$

where $F_{E \times B}$ and $F_{particles}$ are the Poynting and particles fluxes, respectively, and B, ρ and γ are the magnetic field, mass density of particles and the Lorentz factor, respectively. The pulsar wind is assumed to be Poynting flux dominated when it leaves the magnetosphere. Therefore, this magnetization parameter presents values of above 10^4 once the pulsar wind flows out from the *light cylinder*. Nevertheless, observational conditions require a dramatic decrease in this value close to the termination shock. Thus, values of $\sigma \lesssim 0.01$ are needed to explain e.g. the ratio between the synchrotron luminosity and the total spin-down power of the Crab pulsar (Kennel and Coroniti 1984), with Lorentz factor of $\gamma \sim 10^6$, much higher than values expected from free expanding winds (Arons 2002). The reason that sets off this change in the nature of the pulsar wind, from Poynting to particle dominated wind, is not well understood yet (Arons 2009), although magnetic reconnection seems to solve the problem (Porth et al. 2013).

7.2 Pulsar Wind Nebulae

In the termination shock, particles are accelerated with a power-law distribution up to multi-TeV energies. When these particles flow out from the termination shock, they interact with the ISM, giving rise to a PWN. These particles emit synchrotron radiation (due to the existing magnetic field in the PWN), creating a *synchrotron nebula* right after the termination shock, which is detected from radio to X-rays (see Fig. 7.4). The size of this *synchrotron nebula* is inversely proportional to the energy, as more energetic particles loss their energy before traveling long distances. This effect is significant under high magnetic fields, as it is the case for the Crab Nebula. In cases with low magnetic field, as e.g. 3C 58, radio and X-ray nebulae are indistinguishable.

The synchrotron radio emission is characterized by a power-law function defined as $S_\nu \propto \nu^\alpha$, where S_ν is the observed flux density at a frequency ν and α is the spectral index, which ranges from -0.3 to 0. The radio luminosity is $\sim 10^{34}$ erg s^{-1} (between 100 MHz and 100 GHz). The X-ray emission is as well defined by a power-law

Fig. 7.4 Acceleration mechanisms and production sites inside a PWN. Credit: Aharonian (2004)

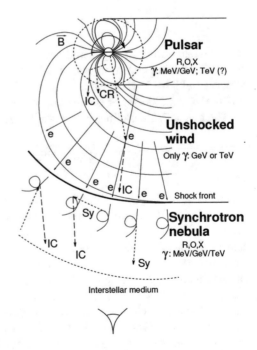

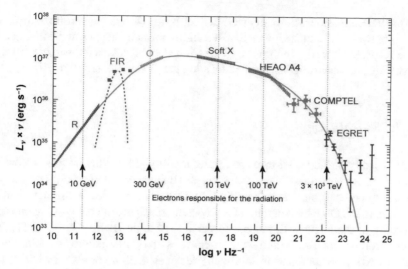

Fig. 7.5 Overall Crab Nebula's synchrotron spectrum from radio to soft gamma rays. The grey solid line represents the fit to the synchrotron emission, whilst the magenta dotted line corresponds to the FIR excess produced by galactic dust. Credit: Gaensler and Slane (2006)

function, $F \propto E^{-\Gamma}$, with a photon index $\Gamma \sim 2$, reaching a luminosity of $\sim 10^{35}$ erg s^{-1} (between 0.5–10 keV). In between, at IR energies, the luminosity is dominated by a component originated by the dust emission that embedded the contribution of the synchrotron IR, as shown in Fig. 7.5 for the Crab Nebula.

7.2.1 Gamma-Ray Emission

The synchrotron radiation extends up to hard X-rays or even soft gamma rays. At higher energies, this process is not efficient anymore and HE and VHE gamma rays are produced instead by UHE electrons via IC scattering on different photon fields. These targets are:

- **CMB**: This photon field is uniformly distributed in the Galaxy and does it all over the entire nebula as well.
- **Far Infrared (FIR)**: Target provided by the excess of photons coming from galactic dust. It could explain up to $\sim 50\%$ of the gamma-ray flux (de Jager and Harding 1992).
- **Synchrotron X-rays**: This photon field gives rise to gamma rays through SSC by the same electrons that produced them. Its contribution is important in young PWNe and dominant in the case of the Crab Nebula.

This VHE emission is normally well-defined by a power-law spectrum with photon indices between 1.3 and 2.8, given that detections normally happen at the falling edge of the IC peak. However, in the case of the Crab Nebula, detected down to ~50 GeV by MAGIC, its spectrum is described by a log parabola (Aleksić et al. 2016).

7.2.2 Morphology

PWNe usually present a jet-torus structure: the torus-like structure arises from the emission at the equatorial plane of the pulsar (Bucciantini et al. 2006) and two jets expand along the toroidal axis (see Fig. 7.6). One of the two jets is brighter than the other due to the Doppler beaming effect, which affects the intensity of the emission according to the angle with respect to our line of sight (Pelling et al. 1987). The synchrotron nebula also presents the so-called wisps. Their nature is unclear but it could arise from synchrotron instabilities (Hester et al. 2002) or places where the $e^{\pm}$ plasma gets compressed (Spitkovsky and Arons 2004). They can vary in timescales of days.

In certain cases, e.g. the Crab Nebula, there are filamentary structures around the non-thermal optical production sites (see Fig. 7.7). Its origin was speculated to be Rayleigh-Taylor instabilities[1] (Hester et al. 1996).

7.2.3 Evolutionary Phases

As mentioned before, PWNe can be embedded into a SNR, forming a *composite* system, in some cases indistinguishable. The evolution of the SNR can affect the development of the PWN. Thus, we can separate different phases of the PWN:

- **Free expansion**: It happens simultaneously with the first evolutionary phase of the SNR. This phase lasts around thousand years (typically <6 kyr), during which the luminosity of the pulsar is the highest. The escape velocity of the pulsar (<500 km/s, Arzoumanian et al. 2002) is much lower than the expansion velocity of the PWN (~10^3 km/s) and the SNR's, remaining at the center of the nebula.
- **Reverse shock collision**: During the Sedov-Taylor phase of the SNR (second phase, see Reynolds 2008), a reverse shock is formed and starts moving inwards in the direction of the pulsar. At some point, it collides with the expanding PWN. The reverse shock compresses the PWN considerably, increasing this way its pressure. This results in a sudden expansion of the PWN. This process is recurrent, causing an oscillation of the nebula that lasts some tens of thousand years. These collisions can produce Rayleigh-Taylor instabilities at the edge of the PWN. Moreover, it

---

[1] The **Rayleigh-Taylor instabilities** are those produced between two fluids of different densities in which the lighter pushes the heavier fluid.

Fig. 7.6 X-ray image of
Crab Nebula taken with
Chandra. The internal
structures of jets, torus and
wisps are depicted. Credit:
NASA/CXC/SAO

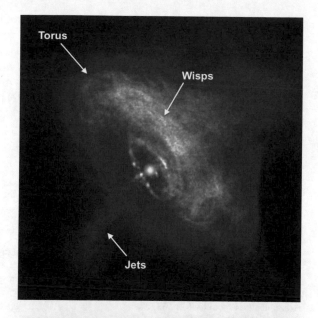

Fig. 7.7 Optical image of
Crab Nebula taken with the
Hubble telescope, where the
filamentary structure is
observed. Image taken from
Hester (2008)

can affect the overall morphology of the PWN and even displace it from the initial pulsar position.

- **Relic phase**: At this last phase, the pulsar has escaped a distance similar or larger than the PWN radius from the initial position and wind bubble. Thus, it leaves a *relic* PWN behind, whilst a new smaller PWN is created at the new position. Without compact object that powers it, the *relic* PWN is usually dominated by IC emission.

References

Aharonian FA (2004) Very high energy cosmic gamma radiation: a crucial window on the extreme Universe (World Scientific Publishing Co)

Aleksić J et al (2016) Astropart Phys 72:76

Aliu E et al (2008) Science 322:1221

Arons J (2002) In: Slane PO, Gaensler BM (eds) Astronomical society of the pacific conference series, vol 271, Neutron Stars in Supernova Remnants, p 71

Arons J (2009) In: Becker W (ed) Astrophysics and space science library, vol 357, Astrophysics and Space Science Library, p 373

Arzoumanian Z et al (2002) 568, 289

Bucciantini N et al (2006) 368, 1717

de Jager OC et al (1992) 396, 161

Gaensler BM et al (2006) 44, 17

Goldreich P et al (1969) 157, 869

Harding AK (2013) Front Phys 8:679

Hester JJ (2008) 46, 127

Hester JJ et al (1996) 456, 225

Hester JJ et al (2002) 577, L49

Hewish A et al (1968) 217, 709

Kennel CF et al (1984) 283, 710

Manchester RN et al (1977) Pulsars

Ostriker JP et al (1969) 157, 1395

Pacini F (1967) 216, 567

Pelling RM et al (1987) 319, 416

Porth O et al (2013) 431, L48

Reynolds SP (2008) 46, 89

Spitkovsky A et al (2004) 603, 669

Chapter 8
Follow-Up Studies of HAWC Sources

8.1 Introduction

On the 15th of April 2016, the HAWC Collaboration released a preliminary second catalog (2HWC) with 39 TeV sources, from which 19 have no association with any known HE or VHE source (within an angular distance of 0.5°). The data used for the analysis comprises 340 days of observation, between the 26th of November 2014 to the 9th of September 2015. The list of sources was sent under a private Memorandum of Understanding MoU between HAWC and several collaborations, including MAGIC and *Fermi*-LAT. A previous catalog (1HWC) had been reported, under the same Memorandum of Understanding MoU, in April 2015 for the HAWC-111 configuration. In this first survey, only three sources were detected after trials, all associated with known TeV sources (Abeysekara et al. 2016). Not only the increased number of tanks and observation time, but also the improvements in calibration, event reconstruction and likelihood method to obtain the TS maps, gave rise to an increase of the number of detected sources with HAWC.

HAWC analysis was performed by means of a likelihood method, described in (Younk et al. 2015). In this method, a source model needs to be assumed, which is applied to all sources in the sky. The model is characterized by the source geometry and the energy spectrum. For the 2HWC catalog, HAWC performed two approaches: a point-like hypothesis by adopting a spectrum defined with a power-law function of spectral index $\Gamma = 2.7$, and extended source searches assuming the source geometry as an uniform disk of 0.5°, 1° and 2° radii, with a spectrum of index $\Gamma = 2.0$. All 2HWC sources presented a pre-trial significance above 5σ, i.e. TS > 25. The systematic uncertainties for the 2HWC analysis were 0.1° for the sources' position and 50% and 0.2 for the reported flux and photon index, respectively. Detailed information of the published catalog can be found in (Abeysekara et al. 2017).

The 2HWC catalog motivated follow-up studies with the MAGIC telescopes as well as *Fermi*-LAT, whose joint work, along with HAWC, allowed us to obtain a multiwavelength view of interesting candidates, presented in this chapter. The

© Springer Nature Switzerland AG 2018
A. Fernández Barral, *Extreme Particle Acceleration in Microquasar Jets and Pulsar Wind Nebulae with the MAGIC Telescopes*, Springer Theses,
https://doi.org/10.1007/978-3-319-97538-2_8

Table 8.1 Photon index, flux normalization at 7 TeV and energy range for each of the selected sources. Values taken from Abeysekara et al. (2017) (from which only statistical errors are reported), with the exception of the energy range, which was obtained in an afterward dedicated HAWC analysis

	Photon index	Flux normalization $[\times 10^{-15}$ $TeV^{-1}cm^{-2}s^{-1}]$	Energy range [TeV]
2HWC J2006+341	2.64 ± 0.15	9.6 ± 1.9	1–86
2HWC J1907+084*	3.25 ± 0.18	7.3 ± 2.5	0.18–10
2HWC J1852+013*	2.90 ± 0.10	18 ± 2.3	0.4–50

main goal was to focus efforts on the 19 sources with no HE and VHE association, to provide new information of unknown candidates. Thus, after evaluating those new TeV emitters, we selected a short list of three candidates (see Fig. 8.1): 2HWC J2006+341 (RA = 301.55°, Dec = 34.18°), 2HWC J1907+084* (RA = 286.79°, Dec = 8.50°) and 2HWC J1852+013* (RA = 283.01°, Dec = 1.38°), from which I analyzed the two first sources. The three of them were already located in the FoV of other MAGIC observations, allowing us to study the sources without performing new dedicated observations. In turn, the selected candidates seemed to provide suitable scientific interest for all the instruments involved in the project, HAWC, MAGIC and *Fermi*-LAT. All of them were detected during the HAWC point-like search. The corresponding photon index and flux normalization obtained during the HAWC analysis for each case are listed in Table 8.1. An afterward dedicated analysis on these three source, computed by HAWC, provided a specific energy range at which they are detected. The corresponding energy range is as well quoted in Table 8.1.

It is worth mentioning that the likelihood analysis performed by HAWC was not optimized for disentangling regions with multiple sources. Therefore, HAWC skymaps show crowded regions. In order to distinguish possible sources, HAWC classified the candidates as *primary* and *secondary*. The former are local maxima (with TS > 25) separated from the closest local maximum at least $\Delta \left(\sqrt{TS} \right) > 2\sigma$. In the latter, this distance is narrower, $1\sigma < \Delta \left(\sqrt{TS} \right) < 2\sigma$, and therefore they could be non-independent sources but part of *primary* ones. The *secondary* sources are labeled with an asterisk at the end of the name, as it is actually the case for 2HWC J1907+084* and 2HWC J1852+013*.

8.2 Observations and Data Analysis

As mentioned above, the three sources were situated in the FoV of previous MAGIC observations and hence, archival data was used for this study. The archival data was taken making used of the *wobble* pointing mode (Sect. 2.4.2.1). This implies that the

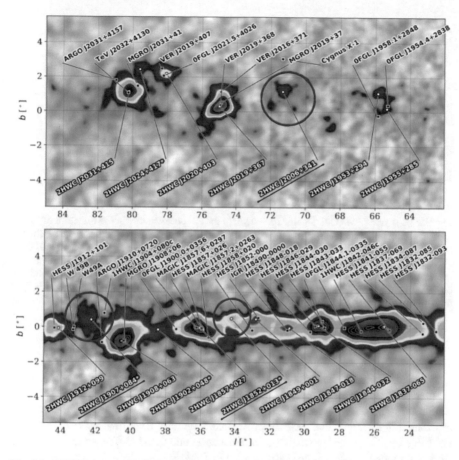

Fig. 8.1 HAWC skymaps of the interesting FoV for the joint project with MAGIC and *Fermi*-LAT, where the position of the three analyzed candidates 2HWC J2006+341, 2HWC J1907+084* and 2HWC J1852+013* is highlighted with blue circles and lines. Skymaps taken from Abeysekara et al. (2017)

coordinates of the HAWC sources were shifted from the camera center a different distance than the standard offset of 0.4° (see Fig. 8.2). To account for their location in the camera, the background is evaluated following the OfWP method. Let's assume two wobble positions, W1 and W2, both at 0.4° from the nominal source at opposite directions. For the W1 observations, the OFF is obtained from W2 runs in the same region of the camera in which the source lays in W1. If we have now N wobble pointing (W1, W2, W3...WN), we can obtain the ON region for one of them and N-1 OFF regions from the counterparts. The limitation on the used number of OFF regions, depends on possible sources in the FoV and the extension of the target source. For this study, only one OFF region was used for the analysis of the three sources, taking W1–W2 and W3–W4 pairs. Thus, ON and OFF regions are obtained from the

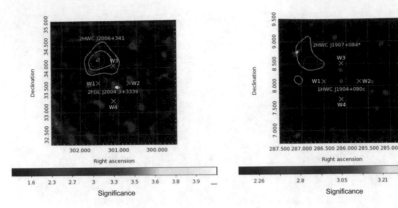

Fig. 8.2 *Left panel*: MAGIC significance skymap for the observations in the direction of 2FGL J2004.3+3339 (cyan diamond). The FoV contains 2HWC J2006+341 located at ∼0.6° from the MAGIC nominal position. *Right panel*: MAGIC significance skymap for the observations of 1HWC J1904+080c (cyan diamond), FoV in which 2HWC J1907+084* is located at 0.79°. The green solid lines correspond in both cases to the HAWC contours. The dashed magenta circle around the position of the 2HWC sources represents the MAGIC extended assumption (radius of 0.16°). The four different *wobble* positions, to which MAGIC pointed during the 2FGL J2004.3+3339 and 1HWC J1904+080c observations, are tagged with W1, W2, W3 and W4 in white color

same part of the camera, accounting this way for the same acceptance and reducing systematic errors in the background estimation. A schematic view of the OfWP method is shown in Fig. 8.3. MAGIC sensitivity depends on the angular offset from the pointing position. Nevertheless, after the upgrade in 2011–2012, the sensitivity at offset angles larger than 0.4° improved considerably as shown by Aleksić et al. (2016b). The distances between the camera center and the HAWC sources at the four different *wobble* positions are summarized in Table 8.2.

The FoV of the two sources I analyzed, 2HWC J2006+341 and 2HWC J1907+084*, were observed under dark and moonlight conditions. The data were divided based on the DC in the MAGIC I telescope, assuming *dark* if $DC_{MI} < 2\,\mu A$, *moderate moon* if $2\,\mu A < DC_{MI} < 4\,\mu A$ and *decent moon* if $4\,\mu A < DC_{MI} < 8\,\mu A$. In the case of *moderate* and *decent moon* samples, artificial noise was added to the MC and background data (used in the RF training) according to the moonlight levels, as described in Sect. 2.4.3.3. The image cleaning values I used correspond to $2–3 \times NSB_{dark}$ and $5–8 \times NSB_{dark}$ levels, respectively (Table 2.3). For the UL calculation on *moon* data, an increased *size* cut according to the strength of the Moon was applied, as shown in Table 2.6.

Taken into account that the three selected candidates for this joint project were detected during the HAWC point-like search, one cannot discard them to be point-like for MAGIC as well. Therefore, for the analysis of 2HWC J2006+341 and 2HWC J1907+084*, I performed two approaches: first, I assumed the sources to be point-like for MAGIC ($PSF_{68} \lesssim 0.10°$ beyond a few hundred GeV), and second, I computed

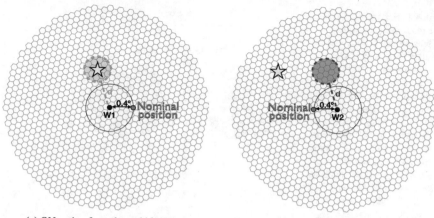

(a) ON region from the wobble W1   (b) OFF region from the wobble partner, W2.

Fig. 8.3 Scheme of the OfWP evaluation background method used in *wobble* pointing mode. The black circle corresponds to the center of the camera, located at 0.4° from the nominal source (gray circle), i.e. the target of the observations. The yellow star represents a source in the FoV that we aim to analyze. While in W1 (*left*) the ON region of the interesting source is selected, W2 (*right*, the wobble partner) is used to get the OFF region at the same position in the camera in which the source stays in W1. The procedure is afterwards performed vice versa, taking ON from W2 and OFF from W1

Table 8.2 Distance in degrees between the four wobble pointing positions (W1, W2, W3 and W4) and the candidates. The observation time, in hours, achieved in each case is also shown

	W1		W2		W3		W4	
	Distance [°]	t_{obs} [hr]	Distance [°]	t_{obs} [hr]	Distance [°]	t_{obs} [hr]	Distance [°]	t_{obs} [hr]
2HWC J2006+341	0.5	16.0	0.9	14.0	0.4	16.3	1.0	14.8
2HWC J1907+084*	0.5	1.0	1.2	1.0	0.7	1.3	1.1	0.9
2HWC J1852+013*	1.1	30.8	0.7	28.8	1.2	29.6	0.6	27.5

the analysis under the hypothesis of extension with radius of $\sim$0.16°. The maximum possible extension is strongly constrained by the standard offset of 0.4° used during *wobble* pointing mode. Taking only one OFF region implies a distance between the center of the position of the source and the background region of 0.8°. The ON region, from which we expect gamma rays, depends, not only on the intrinsic radius of the source, but also on the MAGIC PSF. The latter changes according to the energy, and given that the observations were performed under different moonlight conditions (to which different analysis cuts are applied, see Table 2.3), the energy threshold is not the same for the entire data sample. This ON region is then

Fig. 8.4 Schematic view of
the maximum possible
extension that MAGIC can
assume for a source taken in
wobble mode. Both W1 and
W2 are depicted
simultaneously. From W1,
the ON region (green) is
taken while W2 provides the
OFF region (red) at the same
distance, d, from the camera
center. An intrinsic radius of
the source of ∼0.16° leads to
a θ^2 cut of 0.12 deg^2, i.e.
$\theta = 0.35°$, taken into
account a PSF of 0.07°. The
normalization region is
shown in blue between the
ON and OFF regions

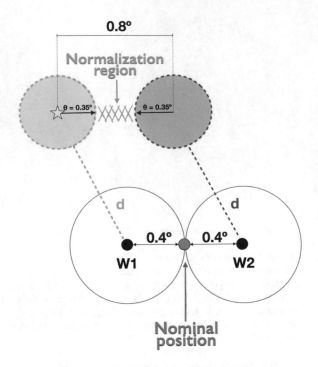

defined by the θ^2 cut applied to compute the flux. It can be described by the equation $\theta^2 \simeq (2 \cdot Radius)^2 + (2 \cdot PSF_{40})^2$, where *Radius* is the assumed radius of
the source, PSF_{40} is the PSF at a 40% containment calculated through one 2-D
Gaussian fit on a Crab Nebula sample at each of the moonlight conditions. To be
conservative, the selected PSF_{40} was the largest one, obtained for the *dark* sample
where lower energies are achieved. This value, $PSF_{40} = 0.07°$, along with the 0.8°
distance between ON and OFF regions, limited the maximum radius to 0.158°, i.e.
$\theta^2 = 0.12$ deg^2 ($\theta = 0.35°$). This way, ON and OFF regions do not overlap, while
in turn there is distance enough between them to perform the normalization between
gamma rays and background events (for a schematic view, see Fig. 8.4).

8.2.1   2HWC J2006+341

2HWC J2006+341 is located at an angular distance of 0.63° from the compact (20")
radio/optical nebula G70.7+1.2, which is embedded in a dense molecular cloud.
G70.7+1.2, at a distance of ∼4.5 kpc from the Earth and situated in the Cygnus
constellation, is a very interesting and unique source due to all the features that
presents:

- It shows a shell-like radio nebula with a broad and blueshifted OI and SII emission lines, which indicate shock with the ISM.
- Millimeter CO emission suggests shock with the molecular material as well.
- A bright NIR Be star is embedded in the nebula.

Given all the aforementioned characteristics, several explanations were proposed to describe its nature: young SNR, nova shell, HII region, Herbig-Haro-like outflows or stellar wind (see e.g., Reich et al. 1985; Green 1986; Jourdain de Muizon et al. 1988; Bally et al. 1989). However, none of these scenarios could explain the low expansion velocities and the non-thermal radio emission detected from its direction. In 1992, (Kulkarni et al. 1992) suggested that the nebula was created by the interaction between the molecular cloud and an hypothetical Be/pulsar binary. This was a speculative theory, whose confirmation depended on the detection of the binary system. With this goal, observations in the X-ray and NIR regimes were carried out. Results with the *Chandra* X-ray Observatory revealed the existence of a point-like source, CXO J200423.4+333907 (see Fig. 8.5), reported by Cameron and Kulkarni (2007). This source was at 3.6" from the previously detected Be star, turning a binary system composed by both of them impossible. However, NIR observations with the Keck Observatory's Laser Guide Star Adaptive Optics (LGS-AO) system showed in turn a relatively bright counterpart around the X-ray source, which was consistent with a highly obscured B-star. This way, (Cameron and Kulkarni 2007) suggested that the X-ray binary (comprised by CXO J200423.4+333907 and a B-star) was the dominant source that powers the radio/optical nebula G70.7+1.2, moving into the molecular cloud from its far side. On the other hand, the Be star could create a reflection nebula on the near side. An unidentified *Fermi*-LAT source included in the Second *Fermi*-LAT catalog (2FGL) catalog, 2FGL J2004.3+3339, was detected at the position of the X-ray source. This discovery reinforced the idea of the binary scenario, although no pulsation was reported so far. In turn, coincident with the *Fermi*-LAT source, a hotspot at the level of 3–4σ was detected on the re-analysis of the 8-years data sample from the Milagro gamma-ray Observatory, whose position was not confirmed by the 2HWC catalog.

MAGIC pointed to the *Fermi*-LAT source and amounted a total of $\sim$55 h (after quality cuts). The data were recorded between April 2015 and August 2016 for a zenith range between 5° and 50°.

8.2.2 2HWC J1907+084*

2HWC J1907+084* is located at 0.79° from 1HWC J1904+080c (RA = 286.1°, Dec = 8.1°). The latter was included in the first HAWC survey (Abeysekara et al. 2016) and previously reported within a list of interesting candidates via private communication to all collaboration HAWC had an MoU with, including MAGIC. The list was composed of 6 new TeV candidates (1 extragalactic and 5 galactic sources) with significance above 4.2σ obtained after 9 months of observations with the HAWC-111

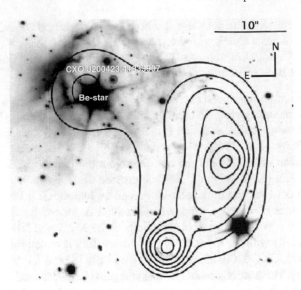

Fig. 8.5 LGS-AO NIR image from G70.7+1.2 with the diffuse *Chandra* X-ray counterparts, taken from Cameron and Kulkarni (2007). The upper left circle corresponds to the X-ray source CXO J200423.4+333907. The Be-star, at 3.6" from the X-ray point-like source, is labeled below. The distance between the expected binary system and the center of the diffuse X-ray emission is around 20"

configuration. The PSF at a 68% radius containment for this HAWC analysis was 0.8° and the systematic uncertainty on the location was expected to be <0.2°. The flux for all candidates above 1 TeV was ∼20% C.U. (with a systematic uncertainty on the flux of ∼40%). The chosen candidate by MAGIC, 1HWC J1904+080c, presented 5.2σ pre-trials, which decreased down to 3.9σ after trials, as shown later on in Abeysekara et al. (2016). The coordinates were not coincident with any known TeV source, but it was close (at 0.3°, coincident within the errors) to a *Fermi*-LAT hotspot (<5σ), 3FGL J1904.9+0818 (Acero et al. 2015).

A point-like source with a flux of ∼20% C.U. in the TeV regime is detectable by MAGIC in less than an hour. However, given the poor HAWC PSF in this analysis, the candidate could be extended and, in turn, presented a high systematic uncertainty on the flux. Thus, it was proposed to observe 1HWC J1904+080c for 5 h, extendable another 5 h if a significance larger than 3σ was reached. Finally, MAGIC performed follow-up observations of the 1HWC candidate from 10th May of 2015 to 19th May 2015, for 6 non-consecutive nights at a medium Zd range of (30, 50)°. After disposing of the data affected by non-optimal weather conditions, the total amount of time reached 4.2 h.

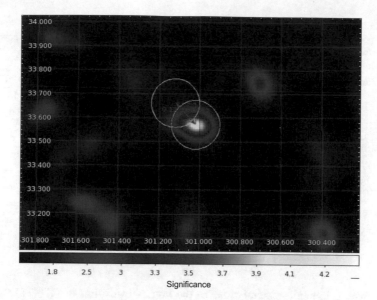

Fig. 8.6 Zoom view of the MAGIC skymap shown in Fig. 8.2 centered in G70.7+1.2 (cyan). The hotspot at the level of 4σ (green) is located at $\sim 0.12°$ from the nominal position. The radius of the circles indicates the MAGIC PSF of $0.10°$

8.3 Results

8.3.1 2HWC J2006+341

No significant signal was found in the direction of 2HWC J2006+341 under any of the assumptions, point-like or extended source. Integral ULs at a 95% C.L. were computed adopting a power-law distribution with photon index $\Gamma = 2.64$, as suggested by HAWC analysis. Accounting for the energy threshold at the maximum moonlight level (*decent moon*, $E_{th} \sim 220 \pm 10$ GeV) and the fact that the sources were detected by HAWC in the TeV regime, integral ULs are given for energies above 300 GeV. For the point-like analysis, this UL was set to 1.8×10^{-12} photons cm^{-2} s^{-1}, while for the extended hypothesis, it increases slightly to 3.8×10^{-12} photons cm^{-2} s^{-1}. Differential ULs for 2HWC J2006+341 were computed as well and can be found in Table 8.3. The integral UL for a point-like source at the position of G70.7+1.2, for energies above 300 GeV and assuming a power-law distribution with $\Gamma = 2.6$, is 1.0×10^{-12} photons cm^{-2} s^{-1}. The left panel of Fig. 8.2 shows the skymap in the FoV of G70.7+1.2, where 2HWC J2006+341 is tagged in green. The skymap is computed assuming a point-like source for a blind scan (without assuming the position of any source). No hotspot arises at the position or nearby 2HWC J2006+341.

Table 8.3 MAGIC 95% C.L. differential flux UL for 2HWC J2006+341 and 2HWC J1907+084* for both point-like ($\lesssim 0.10°$) and extended ($\sim 0.16°$) radius assumptions, considering a power-law spectrum with spectral index of $\Gamma = 2.64$ and $\Gamma = 3.25$, respectively. ULs beyond ~ 4.7 TeV are not computed for J1907+084* due to low statistics

Energy range [GeV]	Differential flux UL for J2006+341 [$\times 10^{-13}$ TeV^{-1} cm^{-2}s^{-1}]		Differential flux UL for J1907+084* [$\times 10^{-13}$ TeV^{-1} cm^{-2}s^{-1}]	
	Point-like	Extended	Point-like	Extended
300–475.5	25.7	45.7	12.1	41.2
475.5–753.6	9.1	61.9	4.8	5.9
753.6–1194.3	2.9	15.0	3.2	3.0
1194.3–1892.9	2.2	7.1	0.6	2.1
1892.9–3000	0.4	1.9	0.2	0.7
3000–4754.7	0.3	0.8	0.04	0.3
4754.7–7535.7	0.1	0.3	–	–

However, a hotspot appears close to the nominal position of G70.7+1.2. With approximately RA = 301.02° and Dec = 33.57°, this hotspot is located at $\sim 0.12°$ from 2FGL J2004.3+3339. It corresponds to point-like emission since it is contained in the PSF of MAGIC (see zoom view in Fig. 8.6). To get an estimation of the significance at this position, again OfWP was used in Odie. In this case, the standard FR cuts were applied, which led to a significance of around 3σ.

8.3.2   2HWC J1907+084*

No excess was found in the direction of 2HWC J1907+084*. The 95% C.L. integral ULs were again computed for energies greater than 300 GeV, but assuming a photon index of 3.25, following HAWC results (see Table 8.1). The integral ULs are 9.7×10^{-13} photons cm^{-2} s^{-1} and 1.4×10^{-12} photons cm^{-2} s^{-1}, for the point-like and radius 0.16° hypothesis, respectively. Under the same conditions, differential ULs were calculated, which are listed in Table 8.3.

1HWC J1904+080c was not detected either. The corresponding integral UL at $E > 300$ GeV, assuming a point-like source defined by a power-law distribution with $\Gamma = 2.6$, is 4.1×10^{-12} photons cm^{-2} s^{-1}. The skymap of the FoV shown in Fig. 8.2 does not reveal any significant emission around the 1HWC or the 2HWC candidates.

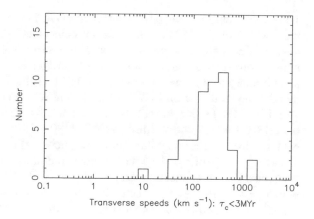

Fig. 8.7 2-D speed of pulsars with a characteristic age of less than 3 Myr. Plot taken from Hobbs et al. (2005)

8.4 Discussion

Given that the largest population of TeV emitter in our Galaxy are PWNe, it would not be improbable that our selected candidates belong to this type of sources. However, the non-detection at lower energies neither with MAGIC or *Fermi*-LAT (see Ahnen et al. 2017a for more information on the *Fermi*-LAT analysis and results), complicates disentangling the nature of these sources. To delve more into the possible PWN nature, I looked for detected pulsars nearby the position of the selected candidates using the Australia Telescope National Facility ATNF catalog[1] (Manchester et al. 2005).

For 2HWC J2006+341, the closest pulsar (and the only one within a 1° radius) is PSR J2004+3429, which lays at ∼0.4° from the 2HWC position. This pulsar is located at a distance of 10 kpc from Earth, with a spin-down power and characteristic age of $\dot{E} = 5.8 \times 10^{35}$ erg s^{-1} and $\tau = 18$ kyr, respectively. Although the spin-down power of the pulsar seems energetic enough to power a TeV PWN (based on observational criterion, see e.g. Collaboration et al. 2017), the distance between PSR J2004+3429 and 2HWC J2006+341 allowed to discard connection between both objects: with an offset of 0.4°, i.e. ∼70 pc, the escape velocity of the pulsar would need to reach ∼4000 km/s. The mean 2D speed for young and old (<3 Myr) pulsars was determined to be 307 ± 47 km/s by Hobbs et al. (2005) in a study involving 233 pulsars. From this sample, the highest 2D speed reached 1624 km/s. Therefore, a pulsar velocity of ∼4000 km/s is unreliably high.

On the other hand, if 2HWC J2006+341 is a point-like source, the distance of 0.63° between this and G70.7+1.2 makes a direct connection between both very unlikely. However, although 2HWC J2006+341 was detected during the point-like search with HAWC, an estimated radius of 0.9° was given in the 2HWC catalog. This value is assumed to be the extension from the source after assigning to it the halo-like structures visible in the residual skymaps. With such extension, 2HWC

---

[1] http://www.atnf.csiro.au/people/pulsar/psrcat/.

J2006+341 could interact or be related with G70.7+1.2 vicinity. This way, although the position of the 2HWC source is not coincident with the one reported for the hotspot of Milagro, TeV emission would be confirmed to take place on this crowded region. Nevertheless, as mentioned in Abeysekara et al. (2017), this radius is just a preliminary result which should not be taken as the definite extension. Dedicated morphology studies with HAWC or CTA would be needed to shed light on this matter.

2HWC J1907+084* presents two pulsars nearby: PSR J1908+0833 at 0.30° from the 2HWC source's coordinates, and PSR J1908+0839, at 0.33°. The former lays at ∼11 kpc from the Earth, with a spin-down power of $\dot{E} = 5.8 \times 10^{32}$ erg s^{-1} and a characteristic age of $\tau = 4.1$ Myr. The poor spin-down power that presents dismisses the probability of connection with a TeV PWN. PSR J1908+0839, at 8.3 kpc and with $\tau = 1.2$ Myr, is more energetic with a spin-down power of $\dot{E} = 1.5 \times 10^{34}$ erg s^{-1}, and therefore a relation between this pulsar and 2HWC J1907+084* cannot be directly excluded. Given the characteristic age of the pulsar, the corresponding PWN would remain in the so-called *relic phase*. This is the last stage of a PWN during which the pulsar escaped from its initial position leaving an old PWN behind with no compact object that injects magnetic or particle flux. Thus, the emission is typically IC-dominated. Therefore, at this stage it is expected to see the pulsar shifted from the PWN position. Following the same exercise as that for 2HWC J2006+341, with a separation of 0.33° in this case (which corresponds to ∼48 pc) and a characteristic age of 1.2 Myr, the obtained escape velocity of the pulsar is ∼40 km/s, which is low but plausible (see Fig. 8.7). Let's assume that PSR J1908+0839 is indeed the pulsar that powers 2HWC J1907+084*. This way, the gamma-ray emission detected by HAWC would be produced by the IC scattering between accelerated electrons, injected by the pulsar, and a low-energy photon field. As shown in Sect. 7.2.1, the target photon field is usually composed by CMB and IR photons. In order to emit gamma rays above $E_\gamma = 200$ GeV (approximately the minimum energy detected by HAWC in this source), the energy of the electron has to be, at least, $E_e \sim 7.5$ TeV, given by $E_e = 17 \cdot (E_\gamma/\text{TeV})^{0.54+0.046 \cdot log(E_\gamma/\text{TeV})}$ (Aharonian 2004).

Making use of the `Naima`[2] software (Zabalza 2015), one can obtain the electron energy distribution necessary to produce the gamma-ray emission from 2HWC J1907+084* (as described by HAWC, see Table 8.1), under certain assumptions. For this calculation, I considered the energy spectrum of the electron population defined by a power law, and IC on CMB and IR photons as the radiative mechanism responsible for the gamma rays. The energy density of the target photon field was set to standard values of $u_{CMB} = 0.25$ eV/cm^3 and $u_{IR} = 0.30$ eV/cm^3, for the CMB and IR radiation, respectively. The obtained electron energy distribution that would be necessary to explain HAWC results on 2HWC J1907+084*, under the aforementioned considerations, is defined by $dN/dE = (5.4^{+1.2}_{-0.8}) \times 10^{49}(E/1\text{TeV})^{-4.69\pm0.15}$ TeV^{-1}. The total energy carried by these electrons for energies above 7.5 TeV is $W_e = (1.23 \pm 0.13) \times 10^{47}$ erg (Fig. 8.8).

On the other hand, the cooling time of electrons in the Klein-Nishima regime can be computed by:

---

[2]http://naima.readthedocs.org.

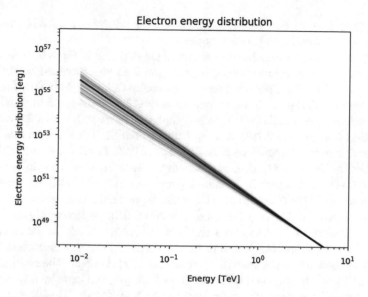

Fig. 8.8 Simulated electron energy distribution for 2HWC J1907+084*, assuming it is powered by the pulsar PSR J1908+0839

$$t_{cool} = 3 \cdot 10^8 \left(\frac{E_e}{\text{GeV}}\right)^{-1} \left(\frac{u}{\text{eV/cm}^3}\right)^{-1} \text{ [yr]} \tag{8.1}$$

where u is the total energy density of the environment (Aharonian 2004). Considering both synchrotron and a Klein-Nishima approximation for IC losses, u can be approximated by the following expression as given by Moderski et al. (2005):

$$u = \frac{B^2}{8\pi} + u_{CMB}\left(1 + 0.01 \cdot \frac{E_e}{\text{TeV}}\right)^{-3/2} + u_{IR}\left(1 + 0.1 \cdot \frac{E_e}{\text{TeV}}\right)^{-3/2} \tag{8.2}$$

where B is the magnetic field. Due to the lack of precise information for 2HWC J1907+084*, I assumed the minimum possible value of B $= 3 \, \mu$G, i.e. the interstellar magnetic field. Again, I assume $u_{CMB} = 0.25$ eV/cm^3 and $u_{IR} = 0.30$ eV/cm^3, as done for the electron energy spectrum computation. These assumptions lead to a $t_{cool} = 7.2 \times 10^4$ years. Given the spin-down power of PSR J1908+0839, $\dot{E} = 1.5 \times 10^{34}$ erg s^{-1}, during a time period equal to t_{cool}, the energy released by the pulsar was $W'_e = \dot{E} \cdot t_{cool} = 3.4 \times 10^{46}$ erg. Therefore, even assuming that all the energy injected by the pulsar, W'_e, was applied on accelerating electrons above 7.5 TeV, the system would not be energetic enough to power a PWN with the gamma-ray emission as that detected by HAWC in 2HWC J1907+084*, which requires $W_e > (1.3 \pm 0.13) \times 10^{47}$ erg. Consequently, the connection between the 2HWC source and PSR J1908+0839 can be discarded. Note that the assumption of the low magnetic field and u_{IR} provides an upper limit on the t_{cool} and so does in the released pulsar energy. Therefore, higher more reliable magnetic field inside the PWN would still lead to the non-connection between objects.

The SED for the two candidates I analyzed are depicted in Fig. 8.9. MAGIC and *Fermi*-LAT analyses were computed with the photon index provided by HAWC in each case (see Table 8.1). The multiwavelength context allow us to obtain information on the candidates' extension. In the case of 2HWC J2006+341, the MAGIC point-like hypothesis (with radius $\lesssim 0.10°$) is completely discarded, whilst the extended assumption (radius of $\sim 0.16°$) is in agreement with the HAWC spectrum within errors, and therefore it could be plausible. For 2HWC J1907+084*, both MAGIC hypothesis (point-like and extended) are compatible with HAWC results at energies above ~ 900 GeV. However, in the sub-TeV regime, neither MAGIC nor *Fermi*-LAT ULs agree with HAWC spectrum. This could be understood in two ways: on one hand, the source could be very extended, well above $0.16°$ which would increase the MAGIC and *Fermi*-LAT ULs considerable above HAWC level. On the other hand, one has to consider that HAWC spectrum is bounded by the energies at which 75% of the events contribute to the total TS in the first and last size bins. The constraint will depend on the declination and spectral index of the source. Consequently, detection along the broad spectrum is not confirmed from HAWC side. Therefore, this mismatch between MAGIC, *Fermi*-LAT and HAWC results could be due to the fact that this candidate does not emit in the energy range covered by MAGIC and *Fermi*-LAT (or presents different spectrum), which is supported by the already very restricting ULs obtained by both instruments.

8.5 Conclusions

MAGIC performed a dedicated analysis on 3 new TeV sources detected on the second catalog of the wide FoV observatory HAWC, of which the results from 2HWC J2006+341 and 2HWC J1907+084* were presented in this chapter. None of them were detected at lower energies and no hotspot was found nearby them. Nevertheless, MAGIC ULs, computed for two hypothesis based on the radius of the source ($\lesssim 0.10°$ and $0.16°$) allowed to constrain the extension of the candidates. The crowded region in which 2HWC J2006+341 is located makes difficult disentangling the extension from the HAWC analysis. However, the assumption of $0.16°$ radius made with MAGIC data seems to be already compatible with HAWC results. In the case of 2HWC J1907+084*, MAGIC and HAWC results are in agreement in the TeV regime. Below 900 GeV, the consistency is not fulfilled anymore, suggesting a larger radius than $0.16°$ or different spectrum in the GeV band. It is worth mentioning that the third candidate included in this project, 2HWC J1908+013*, was not detected either. Integral ULs for a power-law distribution with $\Gamma = 2.90$ were established at 3.8×10^{-13} photons cm^{-2} s^{-1} and 1.7×10^{-12} photons cm^{-2} s^{-1} for the point-like and extended

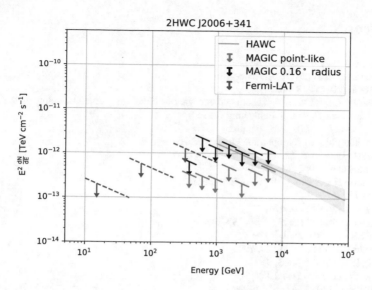

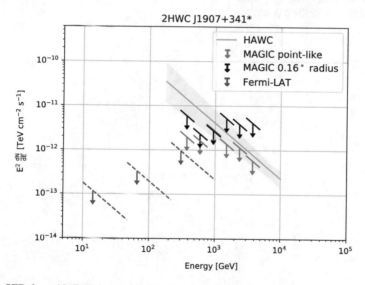

Fig. 8.9 SED from 10 GeV up to ∼90 TeV. In all cases, the assumed spectrum for the sources is a power-law function with photon index $\Gamma = 2.64$ for 2HWC J2006+341 (*top*) and $\Gamma = 3.25$ for 2HWC J1907+084* (*bottom*), as obtained by HAWC (see Table 8.1). *Fermi*-LAT ULs for a point-like assumption are shown in orange. MAGIC results for point-like hypothesis (grey) and 0.16° radius extension (black) are displayed. HAWC butterfly is obtained for the parameters given in Table 8.1

hypothesis, respectively. Finally, it was probed that none of the nearby detected pulsars can be related to the HAWC sources, questioning the possible PWN nature of the VHE emission. In order to reveal more information about the nature of these new TeV emitters, further multiwavelength studies are necessary.

References

Abeysekara A U et al (2016) 817(3)
Abeysekara AU et al (2017). ArXiv e-prints
Acero F et al (2015) 218, 23
Aharonian FA (2004) Very high energy cosmic gamma radiation: a crucial window on the extreme
    Universe. World Scientific Publishing Co
Ahnen ML et al (2017a). (in preparation)
Aleksić J et al (2016b) Astroparticle Physics 72:76
Bally J et al (1989) 338, L65
Cameron PB et al (2007) 665, L135
Collaboration HESS et al (2017). ArXiv e-prints
Green DA (1986) 219, 39P
Hobbs G et al (2005) 360, 974
Jourdain de Muizon, M et al (1988) 193:248
Kulkarni SR et al (1992) 360, 139
Manchester RN et al (2005) 129, 1993
Moderski R et al (2005) 363, 954
Reich W et al (1985) 151, L10
Younk PW et al (2015) ArXiv e-prints
Zabalza V (2015). ArXiv e-prints

Chapter 9
PWN Studies Around High Spin-Down Power *Fermi*-LAT Pulsars

9.1 Introduction

As shown in Chap. 7, pulsars are highly magnetized rotating neutron stars born in SN explosions which are constantly releasing their rotational energy in the form of relativistic Poynting and particle flux, the so-called *pulsar wind*. This wind interacts with the ISM, giving rise to a termination shock in which particles are accelerated. When flowing out, the relativistic particles can in turn interact with the surrounding medium generating a magnetized bubble known as PWN. For the first thousand years, this nebula is mainly synchrotron dominated and detected between radio and X-rays. At higher energies, this mechanism is not efficient anymore and gamma rays are instead produced through IC up-scattering of low-energy photons by ultra-high-energy electrons (see Gaensler and Slane 2006). The common target fields are comprised by CMB, FIR and optical photons. Typically, only pulsars with high spin-down power ($\gtrsim 10^{34}$ erg s^{-1}, based on observational criterion, see e.g. Collaboration et al. 2017) are able to induce prominent TeV PWNe, but the order parameters of the population are not currently understood. PWNe can be surrounded by SNRs, created due to the interaction of the expelled material in the SN and the ISM. These combined systems are known as *composite remnants*, as shown in Chap. 7.

PWNe represent the most numerous population of TeV galactic VHE gamma-ray sources. MAGIC observed and deeply studied the most luminous galactic gamma-ray PWN, the Crab Nebula (Aleksić et al. 2015), and discovered the least luminous one up to now, 3C 58 (Aleksić et al. 2014). These objects were also extensively studied by the southern hemisphere IACT HESS, during their HGPS (Collaboration et al. 2017). The HGPS revealed 14 firmly identified PWNe and 20 candidates at the inner part of our Galaxy. The goal of the MAGIC project developed in this chapter is to prove particle acceleration at the outer side of the Galaxy. With this aim, six PWN candidates that host known high spin-down power *Fermi*-LAT pulsars (between $\sim 10^{35}$–10^{37} erg s^{-1}) were selected: PSR J0631+1036, PSR J1954+2838, PSR J1958+2845, PSR J2022+3842, PSR J2111+4606 and PSR J2238+5903. In

© Springer Nature Switzerland AG 2018
A. Fernández Barral, *Extreme Particle Acceleration in Microquasar Jets and Pulsar Wind Nebulae with the MAGIC Telescopes*, Springer Theses, https://doi.org/10.1007/978-3-319-97538-2_9

Table 9.1 Characteristics of the six selected PWNe candidates for the study. *From left to right*: Name of the pulsar, spin-down power, characteristic age, distance and pseudo-distance. The information is taken from the ATNF catalog if not specified otherwise. $\dot{E}/Distance$ is computed using the values from the Distance column when available. In cases where the distance is not well define, as for PSR J2111+4606, pseudo-distance is applied if possible

Name	$\dot{E}$ [erg s^{-1}]	Characteristic age [kyr]	Distance [kpc]	Pseudo-distance [kpc]	$\dot{E}/d^2$ s [erg kpc^{-2} s^{-1}]
PSR J0631+1036	1.7×10^{35}	43.6	2.10	–	3.9×10^{34}
PSR J1954+2838	1.0×10^{36}	69.4	9.2^a	1.6^b	1.18×10^{34}
PSR J1958+2845	3.4×10^{35}	21.7	9.2^a	–	4.0×10^{33}
PSR J2022+3842	3.0×10^{37}	8.9	10	–	3.0×10^{35}
PSR J2111+4606	1.4×10^{36}	17.5	<14.8^c	2.7^b	1.9×10^{35}
PSR J2238+5903	8.9×10^{35}	26.6	<12.4^c	–	5.8×10^{33}

^{a}Tian et al. (2006), ^{b}Saz Parkinson et al. (2010), ^{c}Abdo et al. (2013)

turn, all of them present characteristic age around a few tens of kyr, similar to that displayed by detected PWNe. Basic information from these pulsars, taken from the ATNF pulsar catalogue[1] (Manchester et al. 2005), is summarized in Table 9.1. In cases of radio quiet pulsars, in which distance information is missing, this parameter is taken from the literature. When available, *pseudo-distance* is also provided, which is estimated making used of the spin-down energy loss rate and the gamma-ray luminosity (Saz Parkinson et al. 2010).

9.2   Observations and Data Analysis

From the above mentioned sources, I analyzed all data from PSR J1954+2838, PSR J1958+2845 and PSR J2022+3842 as well as data under dark condition from PSR J2111+4606 and PSR J2238+5903, as presented in the following sections.

The analysis of the data was performed by means of the standard MAGIC analysis software (Sect. 2.4.3). Significance was calculated applying Eq. 2.12. Flux integral and differential ULs were computed following the Rolke method for a 95% C.L., assuming a Gaussian background and a systematic uncertainty of 30% on the effective area.

---

[1]http://www.atnf.csiro.au/research/pulsar/psrcat/.

In all cases, observations were carried out under different moonlight conditions. To perform a proper analysis, the data was divided into three groups according to the mean pixel DC in the camera of MAGIC I during data taking: dark (absence of Moon, $DC_{M1} < 2.0$ μA), moderate moonlight (2.0 μA $< DC_{M1} < 4.0$ μA) and decent moonlight (4.0 μA $< DC_{M1} < 8.0$ μA). The analysis cuts used correspond to those of $1 \times NSB_{dark}$, 2–$3 \times NSB_{dark}$ and 5–$8 \times NSB_{dark}$ (with nominal HV, see Table 2.3), respectively, with the latter slightly modified in terms of pedestal and RMS mean factor to 3.8:1.6 phe in order to mimic more properly the background of the observations. Moreover, appropriated MC-simulated gamma rays and background data (used for the computation of the γ/hadron separation) were necessary for each of the moonlight levels.

9.2.1 PSR J1954+2838 and PSR J1958+2845

These two pulsars, reported in the First *Fermi*-LAT catalog 1FGL (Abdo et al. 2010), are located in a very dense and crowded region, in which structure associations are still under debate (see Fig. 9.1). PSR J1954+2838 is positionally coincident with SNR G65.1+06. The latter corresponds to a very faint SNR at a distance of 9.2 kpc from Earth with an estimated age between 40–140 kyr (Tian et al. 2006). This SNR seems to be associated with another pulsar in the FoV, PSR J1957+2831, given the compatible distance and kinematic age estimation. In turn, an IR source, IRAS 19520+2759, is detected at the south of the remnant, which was found to be related to CO, H_2O and OH emission lines at a distance similar to SNR G65.1+06, which would suggest interaction with molecular clouds. The re-analysis of the eight-year Milagro data sample at the position of the *Fermi* Bright Sources revealed hints of 4.3σ and 4.0σ in the direction of PSR J1954+2838 and PSR J1958+2845, respectively (Abdo et al. 2009b). This emission may originate from the corresponding PWN or interaction of the SNR and molecular cloud, in the case of PSR J1954+2838. In 2010, MAGIC observed these two pulsars in the stand-alone mode with MAGIC I for ∼25 h, resulting in a non-detection (Aleksić et al. 2010). Nevertheless, the major upgrade between 2011–2012 that both telescopes underwent allowed to improve MAGIC sensitivity with respect to former observations.

In the new campaign, MAGIC observed PSR J1954+2838 from April to November 2015 for a total of ∼16 h of good quality data. These data, that span in a broad zenith range from 5–50°, were taken under dark, moderate and decent moonlight levels. In the case of PSR J1958+2845, only moon data were available. After removing data affected by non-optimal weather conditions, the total amount of effective time reached for the latter was ∼4 h, in a zenith range of 10–40°.

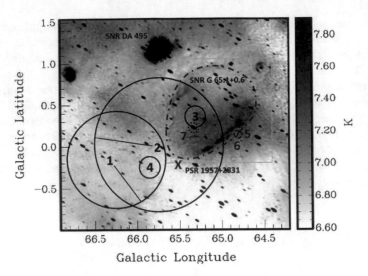

Fig. 9.1 Radio image from the FoV of PSR J1954+2838 and PSR J1958+2845. The numbered objects seen in this skymap are: **a** 3EG J1958+2909; **b** 2CG 065+00; **c** PSR J1954+2838; **d** PSR J1958+2845; **e** region of differing spectral index; **f** IRAS 19520+2759; **g** compact radio object. The errors on the position for (**a**) and (**b**) are shown with solid circles. SNR G65.1+06 is highlighted with dash-dot line. Skymap taken from Tian et al. (2006)

9.2.2   *PSR J2022+3842*

Motivated by the *Chandra* discovery of a point-like source and a faint but distinct surrounding X-ray nebula inside the SNR G76.9+1.0 at 10 kpc, the pulsar PSR J2022+3842 was detected in radio and X-rays (Arzoumanian et al. 2011). Its spin-down power was revised afterwards by Arumugasamy et al. (2014), establishing a value of $\dot{E} = 3 \times 10^{37}$ erg s^{-1}. Therefore, PSR J2022+3842 is one of the highest spin-down power pulsars in the sky. Nevertheless, even though it is extraordinarily powerful, it does not show a bright X-ray nebula and presents, in turn, an unusually low conversion efficiency of spin-down power to X-ray luminosity, as reported by Arzoumanian et al. (2011). These features would imply low magnetization in the medium, that favors the scenario in which most of the rotational energy is converted into IC emission, turning PSR J2022+3842 into one of the best candidates for VHE gamma-ray searches. The source was observed by VERITAS during the Cygnus Survey for 10 h, which ended up in a 3σ UL for energies greater than 200 GeV and a flux at ~3% C.U.. The 8-years Milagro skymaps reported a hint at the 3.5σ level in the region at multi-TeV energies, although no precise flux or energy information on this source is available (Abdo et al. 2009b). MAGIC observed PSR J2022+3842 for ~44 h, covering a large zenith range from 10° to 50°. Given the Milagro hotspot, this source is expected to emit at TeV energies which allowed us to observe it under the all aforementioned moonlight levels.

9.2.3   PSR J2111+4606 and PSR J2238+5903

PSR J2111+4606 and PSR J2238+5903 have a similar high spin-down power around 10^{36} erg s^{-1}. They were both detected during blind *Fermi*-LAT pulsar searches (Pletsch et al. 2012; Abdo et al. 2009a, respectively). The former is located at a distance of 2.7 kpc near the Galactic plane in a low radio emission area, with faint 21 cm structures as shown by The Effelsberg-Bonn H$_I$ Survey (Winkel et al. 2016). On the other hand, PSR J2238+5903 is also located in the Galactic plane although its distance has not been yet constrained. During a search of the 2FGL sources (Nolan et al. 2012), Milagro reported an evidence of emission at the level of 4.5σ at the position of PSR J2238+5903 (Abdo et al. 2014). The PWNe associated to these *Fermi*-LAT pulsars are assumed to be extended in this study, based on former internal MAGIC correlations between surface brightness and flux obtained for the first proposal of observation. Thus, the nebula hosting PSR J2111+4606 is expected to have a radius of $0.15°$, while that surrounding PSR J2238+5903 would display a slightly smaller one of $0.10°$. MAGIC carried out a deep campaign on these sources, accumulating ~55 and ~44 h of optimal quality data on PSR J2111+4606 and PSR J2238+5903, respectively. Under all moonlight levels as well, PSR J2111+4606 was observed for a zenith range of $(5, 50)°$, whilst PSR J2238+5903 data was taken at medium zenith angle, between $30°$ and $50°$.

9.3   Results

In this section, I present the integral and differential ULs for each source as well as the skymaps for interesting FoVs. It is worth mentioning that no significant signal was found for the sixth candidate included in this project, PSR J0631+1036.

9.3.1   PSR J1954+2838 and PSR J1958+2845

No gamma-ray excess was found in the direction of either PSR J1954+2838 or PSR J1958+2845. The measured signal is compatible with background at energies greater than 300 GeV and 1 TeV (the latter motivated by Milagro hotspots). Nevertheless, it is worth stressing that PSR J1958+2845 showed a significance of $\sim2.30\sigma$ (for $E > 1$ TeV) after only ~4 h of observations. No hotspots are highlighted in the significance skymaps from PSR J1958+2845. In the case of PSR J1954+2838, a small hotspot situated at an offset of $\sim0.23°$ from the nominal source appears at the level of $\sim3.5\sigma$, although its position is not coincident with any known system (see Fig. 9.2).

The corresponding integral ULs for energies above 300 GeV assuming a power-law distribution with photon index $\Gamma = 2.6$ are 1.1×10^{-12} ph cm^{-2} s^{-1} ($\sim0.8\%$ C.U.) and 2.5×10^{-12} ph cm^{-2}s^{-1} ($\sim1.9\%$ C.U.) for PSR J1954+2838 and PSR

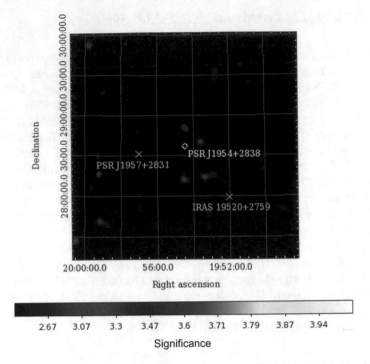

Fig. 9.2 MAGIC significance skymap for the observations of PSR J1954+2838 (white diamond). The pulsar PSR J1957+2831 associated to the SNR G65.1+06 is marked in blue, while the IR source, IRAS 19520+2759, located at the south of the remnant, is shown in green

Table 9.2 MAGIC 95% C.L. differential flux ULs for PSR J1954+2838 and PSR J1958+2845 assuming a power-law spectrum with spectral index of $\Gamma = 2.6$

Energy range [GeV]	Differential flux ULs for PSR J1954+2838 [$\times 10^{-13}$ TeV^{-1}cm^{-2}s^{-1}]	Differential flux ULs for PSR J1958+2845 [$\times 10^{-13}$ TeV^{-1}cm^{-2}s^{-1}]
300–475.5	51.2	134.7
475.5–753.6	15.3	26.9
753.6–1194.3	5.7	5.7
1194.3–1892.9	2.8	5.6
1892.9–3000	4.5	2.6
3000–4754.7	0.6	0.6

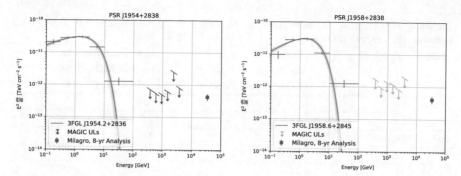

Fig. 9.3 SED for the observation on PSR J1954+2838 and PSR J1958+2845, including results on *Fermi*-LAT pulsar as well as Milagro re-analysis, taken from Abdo et al. (2010, 2009b), respectively

Table 9.3 MAGIC 95% C.L. differential flux ULs for PSR J2022+3842 assuming a power-law spectrum with spectral index of $\Gamma = 2.6$

Energy range [GeV]	Differential flux ULs for PSR J2022+3842 $[\times 10^{-13} \text{ TeV}^{-1}\text{cm}^{-2}\text{s}^{-1}]$
300–475.5	70.8
475.5–753.6	15.5
753.6–1194.3	2.9
1194.3–1892.9	1.3
1892.9–3000	3.6
3000–4754.7	0.6

J1958+2845, respectively. Differential ULs are also listed in Table 9.2, under the same conditions. The SED is depicted in Fig. 9.3, including the *Fermi*-LAT spectrum for the pulsar and Milagro results.

9.3.2 PSR J2022+3842

After ∼44 h of observation, no significant excess was found in the direction of PSR J2022+3842. The FoV in the skymap is compatible with background with no structure or hotspot popping-up nearby the most energetic pulsar of this study. Integral and differential ULs were therefore computed assuming a power-law function of $\Gamma = 2.6$ at 95% C.L. An integral UL for energies greater than 300 GeV is set to 1.5×10^{-12} ph cm^{-2} s^{-1}, which corresponds to ∼1.2% C.U. Differential ULs are quoted in Table 9.3.

Table 9.4 MAGIC 95% C.L. differential flux ULs for PSR J2111+4606 and PSR J2238+5903 assuming a power-law spectrum with spectral index of $\Gamma = 2.6$

Energy range [GeV]	Differential flux ULs for PSR J2111+4606 [$\times 10^{-13}$ TeV^{-1}cm^{-2}s^{-1}]	Differential flux ULs for PSR J2238+5903 [$\times 10^{-13}$ TeV^{-1}cm^{-2}s^{-1}]
300–475.5	106.8	50.2
475.5–753.6	12.5	9.4
753.6–1194.3	2.9	2.9
1194.3–1892.9	1.6	1.5
1892.9–3000	0.6	0.4
3000–4754.7	0.2	0.3

9.3.3 PSR J2111+4606 and PSR J2238+5903

The MAGIC analysis performed, assuming radii of 0.15° and 0.10° for PSR J2111+4606 and PSR J2238+5903, respectively, yielded no detection. As done for previous sources, we computed integral ULs above 300 GeV, considering a standard power-law distribution with photon index 2.6. The constraining ULs are 1.4×10^{-12} ph cm^{-2} s^{-1}, i.e. 1.1% C.U., in the case of PSR J2111+4606, and 8.9×10^{-13} ph cm^{-2}s^{-1}, which corresponds to 0.7% for PSR J2238+5903 (Table 9.2).

9.4 Discussion and Conclusions

Despite observing PWNe hosting pulsars with spin-down power as large as the one for PWNe already detected at VHE, no detection was achieved for any of the six PWN candidates selected for this project and, in particular, the ones included in this thesis. Most of them, except for PSR J1958+2845, were observed for a large amount of hours, allowing to explore flux levels below what had been detected from other PWNe. Given that our observations were performed in the outer parts of the Galaxy, a non-detection could be explained if different behaviors are found in the MAGIC candidates with respect to those shown by detected PWNe located in the inner regions. To highlight any possible difference, I compare in Fig. 9.4 the PWN luminosity between 1 and 10 TeV of the five PWNe shown in this chapter with respect to the characteristic age and spin-down power of the hosted pulsars. In turn, I included all detected PWNe (inside and outside of the HGPS), along with the HGPS candidates and the ULs obtained for the undetected HGPS PWNe (see Collaboration et al. 2017). The luminosity and gamma-ray efficiency conversion of the five above mentioned PWNe are also quoted in Table 9.5.

MAGIC results are in agreement with the overall behavior observed by HESS using detected TeV PWNe and ULs. Therefore, one can conclude that the PWNe

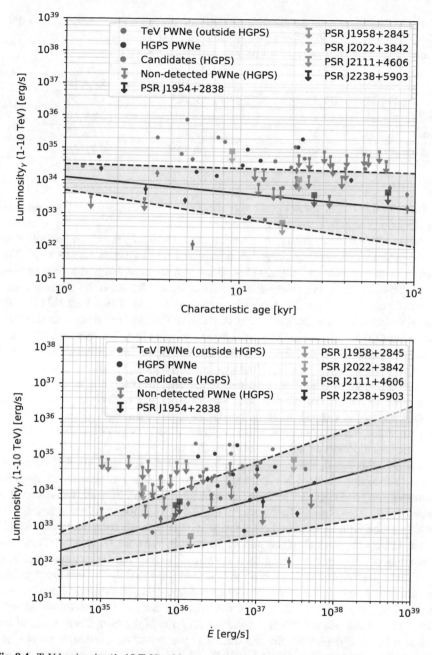

Fig. 9.4 TeV luminosity (1–10 TeV) with respect to the characteristic age (*top*) and the spin-down power of the pulsar (*bottom*). The five candidates analyzed in this thesis are marked with squares, while external PWNe are shown with circles. In the latter, detected PWNe from inside and outside of the HGPS, candidates and non-detected nebulae from it are included. The fit obtained in the study of PWNe by Collaboration et al. (2017) is depicted as a blue band

Table 9.5 *From left to right:* Integral UL above 300 GeV, UL on the TeV luminosity (1–10 TeV), efficiency converting rotational energy into TeV gamma rays ($L_{\gamma,1-10TeV}/\dot{E}$)

Name	Integral UL [$\times 10^{-12}$ ph cm^{-2} s^{-1}]	$L_{\gamma,1-10TeV}$ [erg s^{-1}]	ξ
PSR J1954+2838	1.1	5.2×10^{33}	5.2×10^{-3}
PSR J1958+2845	2.5	1.2×10^{34}	3.5×10^{-2}
PSR J2022+3842	1.5	8.4×10^{34}	2.8×10^{-4}
PSR J2111+4606	1.4	5.7×10^{32}	4.1×10^{-4}
PSR J2238+5903	0.89	7.7×10^{33}	8.7×10^{-3}

included in this thesis are not outliers of the TeV PWN population and so, gamma-ray emission could be expected given their features.

The possible reasons for a non-detection can be basically encompassed into two: an extension of the sources larger than expected or low target photon field. The former would applied to PSR J2111+4606, given its most likely distance or even PSR J1954+2838, if its *pseudo-distance* is finally confirmed to be more accurate than its distance of 9.2 kpc. For the analysis presented in this thesis, PSR J2111+4606 was already assumed to have a 0.15° radius. Nevertheless, for completeness, larger extension of 0.2–0.3° for the closest sources included in this project are currently being analyzed. Given that these analyses are still preliminary, final results are not shown here, but there is no significant excess or hint for any of the candidates so far. Therefore, this would discard larger extension as the reason for a non-detection, at least up to a radius of 0.2–0.3° that MAGIC can check given the observational conditions. Larger extensions could be studied with new observations using larger *wobble* offset, but without other wavelength information that constrains the radius, any other assumption would be purely speculative.

On the other hand, the scenario of a low target photon medium that decreases the IC interaction chances is reinforced by Fig. 9.5. This plot reveals an apparent accumulation of detected PWNe in the spiral arm named *Scutum*, while no other tendency becomes obvious for the rest of the Galaxy. In order to investigate a speculative relation with the photon field density in each arm, we computed the FIR field in the Galaxy, by means of the publicly available software GalPROP.[2] The program resolves the transport equation of cosmic rays along our Galaxy and calculates the diffuse gamma-ray and synchrotron emission produced during this propagation, assuming certain inputs derivatives of observational data and theoretical predictions. In Fig. 9.6, the obtained FIR photon field density for each aforementioned non-/detected and candidate PWN is shown as a function of the spin-down power of the pulsar it hosts. The plot is split into three regions, based on a tentative probability of detection given the current data at VHE. The division is done searching for the largest fraction of detected (the rightmost line) and non-detected (the leftmost line) sources, forcing that at least a minimum of ten sources are included in each part.

---

[2]https://galprop.stanford.edu/.

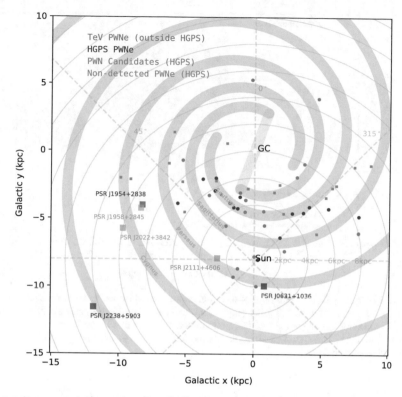

Fig. 9.5 Schematic view of the Galaxy with the spiral arms in which sources included in this study are marked with colored squares. PWNe detected (inside and outside the HGPS) are depicted as orange and dark blue dots, while candidates included in the HGPS are shown as light blue dots. Pulsars with spin-down power above 10^{35} erg s^{-1} for which non-detection of PWN was achieved during the HGPS are also represented as grey squares. The spiral arms were constructed making use of the open source gammapy python packages (http://docs.gammapy.org)

For the division, the PWN candidates, whose nature is uncertain, were not used. The plot evidences a correlation between these two features: most of the detected PWNe present high spin-down power, while surrounded in turn by a large FIR energy density photon field. This way, it is probed that not only a high $\dot{E}$ is necessary ($\gtrsim 10^{36}$ erg s^{-1}), but a large target photon field would increase the probability of detection. The features of the five PWNe I analyzed place them into the region of the plot where the percentage of non-detected sources is as high as 95%, with the only exception of PSR J2022+3842. The latter is confined in the middle region and hence, the probability of detection could be higher. Nevertheless, despite its high spin-down power, this PWN shows a very low FIR field. One can compare this source with a similar detected PWN, e.g. G0.9+0.1. Both systems are located at alike distances (10 and 13 kpc, respectively) and powered by pulsars with similar spin-down power (PSR J2022+3842 with $\dot{E} = 3. \times 10^{37}$ erg s^{-1} and PSR 1747-2809, hosted by G0.9+0.1,

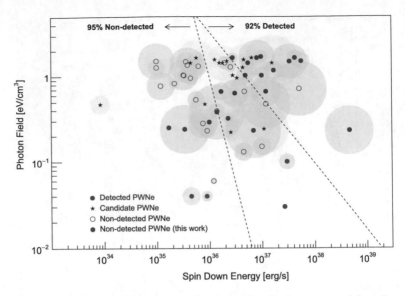

Fig. 9.6 FIR energy density photon field as a function of the spin-down power of the pulsars. The detected (red dots), non-detected (blue empty dots) and candidate PWNe (grey stars) included in the HGPS are shown, along with MAGIC candidate sources (blue filled dots) included in this study. The two black dashed lines evidence that most of the detected PWNe present high spin-down power and high FIR photon field

with $\dot{E} = 4 \times 10^{37}$ erg s^{-1}), but with opposite results in the TeV regime. According to former standards, the fact that one source is detected whilst the other is not, even having similar spin-down power and luminosity, could be understood as a contradictory result. Nevertheless, it can be explained attending to the tendency between photon field and spin-down power exposed in Fig. 9.6. As shown in this plot, our candidate presents a very low surrounding FIR photon field of 0.10 eV/cm^3, compared to the high energy density of 1.64 eV/cm^3 displayed by G0.9+0.1. Therefore, this correlation could explain the MAGIC non-detection on these five candidates. In order to disprove that this tendency is biased by observations, the luminosity of each source is included as semi-transparent colored circles around each dot: the larger circle, the more luminous the object is. Thus, it is demonstrated that not only PWN with high luminosity are detected or follow the correlation.

These results on galactic PWNe are also in agreement with the scenario proposed for N157B, the extragalactic PWN located in the Large Magellanic Cloud and powered by the most powerful known pulsar (with $\dot{E} = 4.9 \times 10^{38}$ erg s^{-1}). Although it is not an efficient synchrotron accelerator compared to the Crab Nebula, given its low magnetic field, it was probed to be a much more efficient gamma-ray emitter, which was suggested to be related with an increased photon field. Finally, it is worth pointing out that the two candidates, J1745-303 and J1746-308, reported by H.E.S.S and associated with the ATNF pulsar B1742-30 (with $\dot{E} \sim 8.5 \times 10^{33}$ erg s^{-1}) behave as outliers within the entire population, which could point to a different nature rather than PWN to the VHE emission or a different pulsar powering them.

References

Abdo AA et al (2009a) Science 325:840
Abdo AA et al (2009b) 700, L127
Abdo AA et al (2014) Astroparticle Physics 57:16
Abdo AA et al (2010) 188, 405
Abdo AA et al (2013) 208, 17
Aleksić J et al (2015) Journal of High Energy Astrophysics 5:30
Aleksić J et al (2010) 725, 1629
Aleksić J et al (2014) 567, L8
Arumugasamy P et al (2014) 790, 103
Arzoumanian Z et al (2011) 739, 39
Collaboration HESS et al (2017). ArXiv e-prints
Gaensler BM et al (2006) 44, 17
Manchester RN et al (2005) 129, 1993
Nolan PL et al (2012) 199, 31
Pletsch HJ et al (2012) 744, 105
Saz Parkinson PM et al (2010) 725, 571
Tian WW et al (2006) 455, 1053
Winkel B et al (2016) 585, A41

Part IV
LST Camera

Fig. IV.1 L0+L1 mezzanine designed at IFAE. Credit: Scott Griffiths

Chapter 10
Quality Control of LST Camera Subsystems

10.1 Introduction

CTA intends to reach up to 10 times better sensitivity (around 1 TeV) than the current IACTs and extend the energy range. The LST are the responsible for the sensitivity below ~100 GeV and aim to lower the threshold as much as possible. One of the most important systems to success on this purpose is the camera. The reliability of its components depends on the results obtained from the characterization, i.e. dedicated tests used to evaluate the functionality of the devices and check, in this way, if they reach the desired specifications to achieve the goals of the LST.

Along my thesis period, I was involved on the characterization for several subsystems that composed the future LST prototype, among which the PMTs, the Power Supply Units (PSUs) and the L0+L1 trigger mezzanines stand out. In order to set these subsystems in context, in the first section of this chapter I give an overview of the LST camera and its components, including their functionality and information on the L0+L1 trigger mezzanine tests. The tests performed on the PSUs are shown in Sect. 10.3.

10.2 Overview of the LST Camera

Fig. 10.1 shows the electrical diagram of the LST camera. The different cabling needed to connect and power all subsystems is classified in seven types, labeled in the image. The camera itself can be split into three sections: *front*, *middle* and *back*.

© Springer Nature Switzerland AG 2018
A. Fernández Barral, *Extreme Particle Acceleration in Microquasar Jets and Pulsar Wind Nebulae with the MAGIC Telescopes*, Springer Theses,
https://doi.org/10.1007/978-3-319-97538-2_10

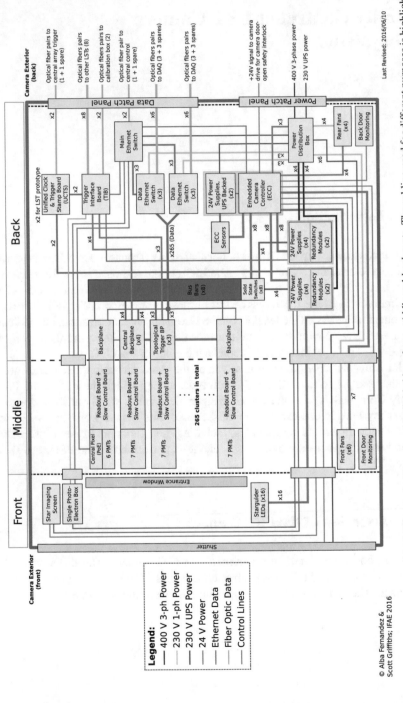

Fig. 10.1 Electrical diagram of the LST camera. The camera is divided into *front*, *middle* and *back* parts. The cabling used for different purposes is highlighted with different colors and its functionality is shown in the legend

10.2.1 Front Part of the LST Camera

This is the outer part of the camera, the one in contact with the environment. The most external component is the shutter or lid, which prevents light from entering in the camera, protecting this way the PMTs from the daylight. In addition, it provides protection against adverse environmental conditions, such as rain, during the day. The goal of the entrance window, placed after the shutter, is to protect the instrumentation from any aggression (rain, dust, e.g.) during observational time. In this *front* side, the starguider LED (used to check the pointing of the camera as done in MAGIC, see Sect. 2.4.1.1), the star imaging screen and the (SPE) box are also installed. The latter works as an additional method that could be used for the calibration of the camera (to check e.g. gain or linearity) and it will be probably not installed in the first LST prototype.

10.2.2 Middle Part of the LST Camera

The main body of the camera concentrates in this region. Besides the front fans, designed to cool the system during operation, the *middle* part of the camera encompasses the 265 clusters, where the trigger and readout take place. Figure 10.2 shows a cluster, where all parts are identified. Each cluster is formed by 7 Hamamatsu PMTs, to which light guides are attached in order to collect the highest amount of photons reflected in the mirrors and excludes, in turn, NSB light coming at larger angles (see Fig. 10.3). The PMTs work with a HV of $\sim$1.4 kV supplied by a Cockcroft-Walton (CW) power supply that is shipped with the PMTs. The output signal of each PMT is preamplified with low noise by the Pre-Amplifier for the Cherenkov Telescope Array (PACTA), an Application-Specific Integrated Circuit (ASIC) designed by the ICC-UB group. Two differential paths form the output of PACTA: the low-gain and the high-gain branch. Having these two paths, a larger dynamic range is possible: linearity in the low gain reaches higher ranges (up to thousand photoelectrons), while the high gain extends linearity at the smallest ones. The PMTs are connected to the readout board through the Slow Control Boards. This device can control the HV of the detectors and monitors different parameters as the DC current, temperature or humidity. Thus, the signal from the seven channels (0–6, one for each PMT) enters in the readout board, which in the case of the LST is called *Dragon*. Each signal is replicated to reach both readout and trigger subsystems. To the trigger board (L0+L1 mezzanine) only a copy of the high-gain paths are sent. The L0+L1 mezzanine is an unique Printed Circuit Board (PCB) board composed by the connectors and the corresponding L0 and L1 ASIC, on charge of the main functionality (as accomodate the signal and take the trigger decisions). After entering in the mezzanine, the signals pass through the delay lines, used to compensate the different transit times and synchronize the PMT signal (within a range of 0–5.75 ns). Thus, the signal reaches the L0 ASIC, first step of the trigger system in which individual pixel signals are

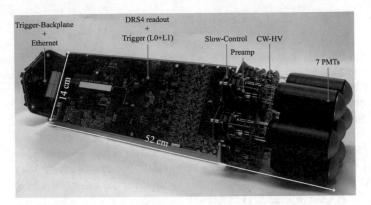

Fig. 10.2 LST module with all components highlighted: 7 photodetectors, followed by the HV chip and the preamplifier (PACTA), the Slow Control board, *Dragon* board in which L0+L1 mezzanine is attached (the trigger board is connected to the opposite side of the board seen in the image) and finally the backplane where the Ethernet switches are placed

evaluated. All signals go through an attenuator in order to compensate for different gains. Here, the voltage can be adjusted between 0.6–1.35 times the input voltage (in steps of 0.05). After the attenuator, two trigger options are available: the sum trigger and the majority trigger.

The sum trigger is the one that will continue to the next L1 level inside the trigger subsystem. The concept is as follows: signals above a certain threshold are clipped to avoid using spurious signals, like APs. There are three clipping options and each of them can be modulated with 63 finer steps. After clipping, signals from the seven channels are summed, which corresponds to the L0 output. On the other hand, the majority trigger discriminates the signals of each pixel according to a specific threshold. If the signal overpasses that threshold, a square 100 mV signal is issued. The emitted signals for each channel are added and the result would constitute the output of the L0. As mentioned before, the majority output is usually disabled and not use in the following steps. Nevertheless, each 100 mV signal is also available as an Low-Voltage Differential Signaling (LVDS) output, from which the L0 Individual Pixel Rate (IPR) is obtained. During this thesis, a fast QC of the L0 trigger in 17 mezzanines was performed, making use of an automatic LabView program that modifies the aforementioned parameters (attenuator, clipping, majority discriminator) to cover all possibles values. A proper characterization had been performed before and hence, the goal of these tests was to ensure the functionality before integrating the mezzanines on the camera. To do so, we set low DT to accomplish trigger always. We injected a 200 mV signal with 3 ns Full Width Half Maximum (FWHM) at 10 kHz and checked the IPR in each channel as well as the output of the signal from the sum trigger path. During these tests, the delay lines were also checked. As result, four mezzanines were detected to be non-functional and sent to repair. An afterwards check proved their recover.

Fig. 10.3 LST modules formed by the seven PMTs, the Slow Control board and the *Dragon* board. No trigger mezzanine is connected. The leftmost modules have attached the light guides, although they do not have any reflective foil inside

Following with the trigger scheme, the output of the L0 decision is sent to the L0 fan-out, placed in the backplane board (located already in the *back* part of the camera, see Fig. 10.1). From there, the L0 output is sent to the six neighbouring clusters and, in turn, it receives the L0 signals from those clusters. Thus, these six signals along with the one of the own cluster is sent to the L1 part of the trigger mezzanine. At the L1 level, a selected combination of L0 signals is compared to a threshold. In the L1 ASIC there are available three adders, which can be used to test different geometrical combinations of the clusters, although only the output of one of them will be used later on. If the sum of the signals overcomes the threshold, the L1 ASIC sends an LVDS signal to the L1 distribution subsystem (also placed in the backplane). This signal is then transmitted to the central backplane and from there to the Trigger Interface Board (TIB), also located in the *back* part of the camera and which will be introduced in the next section. A fast QC was also performed on the L1 trigger during this thesis. The test consisted on injecting 1000 events of 300 mV in burst of 12 MHz divided by 10 kHz and scan the signal in each channel and adder from the minimum DT value up to its maximum. We normalized the gain in each channel with respect to a selected reference channel by measuring its value in each channel a priori. The results we wanted to obtain were the DT at which the rate was half its value (i.e. 500 events). The functionality of the L1 in these 17 mezzanines was probed to be correct and compatible with CIEMAT results.

Finally, the TIB sends the final command back to the central backplane and from there to the other backplanes to start the readout at the front-end electronics. The signal from each channel is saved in a buffer formed by four DRS4 chips (with

Fig. 10.4 Metallic structure
holder for the LST modules.
The modules are placed in
each hole. At the center, two
modules are connected,
which have the light guides
attached

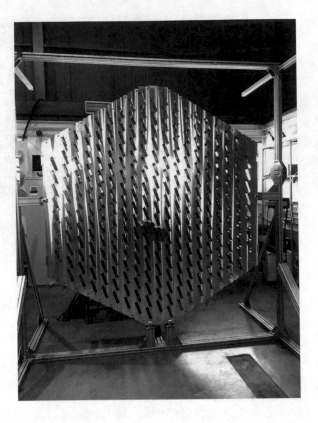

1024 capacitors each) with a longitude of $4\,\mu s$ accounting for a readout speed of
1 GSample/s. The transmission of the data is then performed by Ethernet cables
controlled by a Field Programmable Gate Array (FPGA) situated in the *Dragon*.
All the 265 modules are placed in a metallic structure prepared at CIEMAT (see
Fig. 10.4).

10.2.3   Back Part of the LST Camera

As seen before, the backplanes and the TIB are placed in this section of the camera.
The backplanes are then the connection between the modules and the rest of the
subsystems, a part from connecting all modules together: both the 24 V DC PSU
and the Ethernet connection are provided through the backplanes. The power supply
is distributed owing to eight bus bars. The 24 V line is provided by eight PULS
QT40.241 PSU at 40 A (i.e. 960 W per PSU), as the ones shown in Fig. 10.5. Among
them, there are redundancy modules decoupled to each pair which provide current

Fig. 10.5   PSU and redundancy modules used in the LST camera

in case of a failure in any PSU. More information and dedicated characterization of these PSU are presented in the next section.

Besides these eigth PSU, in the *back* part there are also two 24 V PSU Uninterruptible power supply (UPS) -backed, connected to critical devices for the safety and operation of the telescope. These devices are the Embedded Camera Controller (ECC), main brain of the camera, and the ECC sensors. The main aims of these subsystems are to monitor the environment (like temperature or humidity), control auxiliary components (like fans, starguider LED or the lids) and make decisions to ensure the safety of the camera. ECC does not control the modules or the TIB, but it can get information from them (as the HV or the temperature) to provide reliable information regarding the status inside the camera.

All the 24 V PSU (including the UPS-backed and also the main Ethernet switch) are powered through the Power Distribution Box (PDB), an electrical cabinet located in the back of the camera (see functional diagram in Fig. 10.6). Its functionality is comprehended in four points:

- **Distribute the incoming 400 V 3-phases power and 230 V UPS-backed single-phase power among the subsystems**: The 400 V 3-phases branch powers most of the camera, including the eight 24 V DC PSU. This branch also supplies the data Ethernet switches and cooling fans. The latter subsystems are powered by 230 V single-phase. It is possible to power these 230 V single-phase devices from the 400 V 3-phases line because the amplitude of each phase is ~230 V. On the other hand, the 230 V single-phase UPS-backed line powers sensitive devices, as the

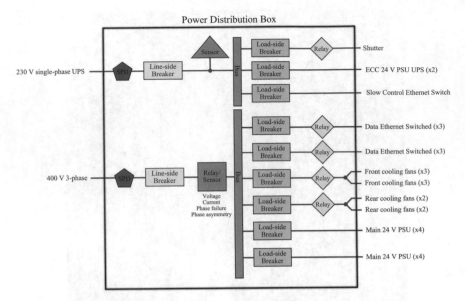

Fig. 10.6 Functional diagram of the PDB. The inputs/outputs are listed along the left/right side of the figure

24 V PSU UPS-backed, the shutter or the main Ethernet switch (the link between the ECC and the outside world).

- **Monitoring the status (voltage, current, etc) of the incoming power lines**: Both 400 and 230 V power lines are connected to sensors in order to track any problem with the incoming power lines. Besides the sensor, the 400 V power line also requires a master control relay, which can be used to cut power in case of maintenance. The 230 V power line does not incorporate such relay in order to avoid shut down the power accidentally.

- **Protecting the camera against electrical transients on the incoming power lines (e.g. lightning)**: This task is carried out internally by Surge Protection Devices (SPDs). Most of the power distribution systems in the LST camera present already some power protection and therefore these SPDs act like an extra protection. On the other side, minor transients in the power lines (like spikes) are handled by the line-side breakers. Finally, the load-side breakers are used to protect the camera subsystems from overcurrents.

- Controlling the flow of power to subsystems, either for safety or for powering them ON/OFF.

All the sensors and relays, along with the SPDs and breakers will be selected at Institut de F'isica d'Altes Energies (IFAE), accounting for the CTA requirements. The mechanical design of the PDB was performed between the Laboratoire Leprince-Ringuet and CIEMAT, while the design and assembly of the PDB internals were carried out by MAES Automation in Sabadell.

10.3 Power Supply Noise Tests

DC power supplies ideally supply a constant voltage or current devoid of time-domain fluctuations. In reality, some amount of noise is always generated or relayed by a power supply. Two types of power supply noise are generally considered, depending on how the noise is transmitted:

- **Conducted emissions**: noise from the power supply is conducted through wires to the device it is powering.
- **Radiated emissions**: electromagnetic radiation emitted by the power supply induces noise in nearby devices.

We further characterize these noise sources in the context of the type of power supply being discussed. We consider two general types of power supplies: linear or switch-mode. In linear power supplies, AC mains voltage at 50 Hz enters the power supply and is rectified to create a positive voltage that fluctuates at a frequency of $\sim$100 Hz. This signal is integrated by large capacitors, regulated to the proper voltage, and filtered to produce a nearly DC signal. A small noise component at 100 Hz, called *ripple*, always survives this process and is included in the output of the power supply. Moreover, higher frequency noise from the mains power may also be present in the output of a linear power supply. Good linear power supplies generally filter both sources of noise very effectively and radiate little, if at all. Switch-mode or switching power supplies have several advantages over linear power supplies, including higher power output and vastly better efficiency, but are generally more noisy. Switching power supplies operate by rapidly switching power transistors on and off at a characteristic *switching frequency*. These supplies generate noise at the switching frequency, or a harmonic thereof, which can be either conducted to the output of the supply or radiated into space. Switching frequencies are generally in the range of 50–5 MHz. Poorly designed power supplies can be a significant source of noise at these frequencies.

The goal of the tests was to characterize the noise generated by the PULS QT40.241 switch-mode power supply, selected to power the LST and NectarCAM cameras. Table 10.1 includes the switching frequencies of the QT40.241. Based on this table, we expect that noise (either conducted or radiated) from the power supply should occur in the 1–300 kHz frequency range, although harmonics could extend this range up to a few MHz.

The PULS QT40.241 will operate in the vicinity of and provide power to PMTs. Noise at the input of the PACTA will directly affect the noise in the output of the PMT, while noise in the HV supply or control lines may affect the stability of the PMT output. It is necessary to be able to detect the electrical signal from SPE emitted by the PMT photocathode. The amplitude of the PMT output signal generated by a SPE depends on the gain of the PMT, but the height of the pulse should have an approximate amplitude of a few mV. It is therefore important to ensure that the noise is well below this level.

Table 10.1 List of PULS QT40.241 switching frequencies

Switching frequency	The power supply has three converters with three different switching frequencies included. One is nearly constant. The others are variable	
Switching frequency 1	105 kHz	Resonant converter, nearly constant
Switching frequency 2	1–150 kHz	Boost converter, load dependent
Switching frequency 3	40–300 kHz	PFC converter, input voltage and load dependent

10.3.1   Equipment and Test Setup

The devices used to perform the tests are the following:

- **PMT**: Hamamatsu R11920-100-20 High QE PMT, serial number ZQ6623. The low voltage input (V_{cc}) supplied to the PMT is +5 V, while the HV control supply is variable over a range of +0.85 V up to +1.5 V and is multiplied by 1000 (yielding an HV range of 0.85–1.5 kV). In the tests that follow, unless otherwise noted, we use a HV of ~1.45 kV, where noise in the PMT should be most prominent and detectable. An SMA connector attached to the base of the PMT module is used for readout.
- **Linear power supply**: Tektronix PS280 A low-noise, dual output laboratory supply is used to supply power to the PMT to establish a baseline. The +5 V and +0.85–1.5 V needed by the PMT can be supplied directly to the module from the power supply. Any laboratory-grade dual output supply can be used for this purpose.
- **Switch-mode power supply**: PULS QT40.241, serial number 11 201 661. *This is the device under test.* It provides a fixed +24 V output, so it cannot directly supply power to the PMT. To power the PMTs with this power supply the Dragon readout board has to be used as an intermediary. The output signal from the PMT can be measured by an oscilloscope or by using the readout board itself.
- **Readout board**: Dragon readout board (v5). The readout board converts the +24V from the PULS QT40.241 to the necessary voltages for the PMT. The readout board can also be used to record signals from the PMTs, which can then be analyzed to search for noise. Communication with the readout board requires an Ethernet connection and Cluster Control (ClusCo) or *rbcp* software.
- **Oscilloscope**: Tektronix TDS3024B digital oscilloscope. A digital oscilloscope is used to view the output signal from the PMT and to perform a Fast Fourier Transform (FFT) of the PMT signal to search for noise in frequency space. The primary reason for using the TDS3024B is its ability to use a 50Ω input impedance (typically oscilloscopes have 1 MΩ input impedance). If using another oscilloscope, ensure correct impedance matching so that feedback oscillations are avoided.

- **Dark box**: Measuring the noise of the PMT output with the HV turned on requires that as little light reaches the PMT as possible. The PMT should be placed in a dark box with panel-mounted cable feedthroughs to avoid exposing the PMT to excess light while it is operating. Note that the dark box may provide some EM shielding depending on its construction. The dark box should be large enough to place the PULS QT40.241 inside, next to the PMT.

10.3.2   Test Description

We evaluate the noise at three different frequency ranges:

- Low frequency: 62.5 Hz, searching for 50 Hz noise (power line noise).
- Medium frequency: 2.5 kHz, searching for noise at the level of kHz (pick-up noise).
- High frequency: Not fixed value, searching for noise at or above 100 MHz.

The results, shown as Voltage versus Time plots, are obtained, in turn, with three different baseline configurations:

- All devices powered off.
- PMT pre-amplifier powered on (at 5 V), PMT HV off.
- PMT pre-amplifier and HV powered on.

The power line (conductive) noise tests were performed using both the linear PSU and PULS QT40.241 PSU. The noise induced by the CTA PSU, using the *Dragon* Board as intermediate, was studied through its trigger path (L1 output). In Fig. 10.7, a scheme of the setup used to carry out these measurements is shown. Two configurations were tested: one with the PMT connected to the Slow Control Board, and the other without PMT.

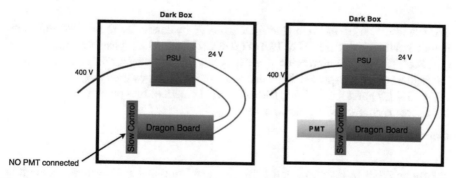

Fig. 10.7 Schemes for conducted noise tests when the PMT is powered by the PULS QT40.241 PSU. To perform these measurements the *Dragon* Board is needed as intermediate between the PSU and the PMT

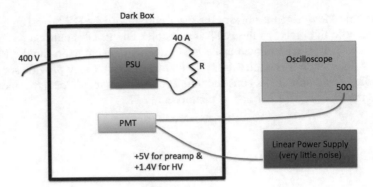

Fig. 10.8 Scheme for radiated emission tests. The signal from the PMT, powered by the linear PSU, is saved by the oscilloscope. The PULS QT40.241 PSU, device under test, is placed close to the PMT inside the dark box

On the other hand, the pick-up (radiative) noise tests were carried out powering the PMT with the linear PSU while the PULS QT40.241 PSU was placed along-side the PMT inside the dark box. The noise was tested under three different test configurations:

- PMT powered off, PULS QT40.241 PSU on (unloaded).
- PMT powered on, PULS QT40.241 PSU on (unloaded).
- PMT powered on, PULS QT40.241 PSU on (loaded).

The load applied to the system was $0.6\,\Omega$ in order to obtain the nominal current of 40 A. The scheme in Fig. 10.8 illustrate this configuration.

10.3.3   Results

The test results allow us to conclude that no noise, by conducted or radiated emission, is induced by the linear PSU or PULS QT40.241 PSU at any possible configuration (see Sect. 10.3.2). For convenience, we only show in this chapter the results at which the highest noise is expected, it means, when the PMT is powered on ($V_{cc} = 5\,V$ and HV $= 1.4\,kV$) and the PULS QT40.241 PSU is loaded. In Appendix C, the results for different configurations are shown.

Conducted noise tests

- **Linear PSU**: In Figs. 10.9 and 10.10, the PMT output signal is shown when it is powered by the linear PSU. The sampling frequency is changed to cover all the above-mentioned range of interest (see Table 10.1).
- **QT40.241 PSU**: The following results were obtained making use of the PULS QT40.241 PSU that powers the readout *Dragon* Board at which the PMT (through

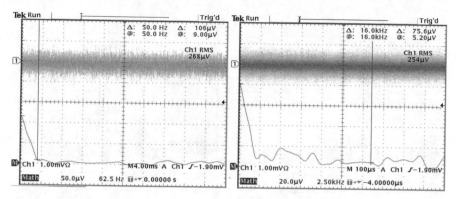

Fig. 10.9 Voltage versus time at different sampling frequency levels of 625 kHz and 25 kHz, respectively, looking for low +and medium frequency noise is shown in the upper part. The vertical scale is 1 mV/div for both cases. With a solid line, at the bottom of each plot, the FFT is shown with a vertical scale of 50 μV/div and 20 μV/div, respectively. For these measurements, all the involved devices were turn on (see Sects. 10.3.2 and 10.3.3)

the Slow Control board) is attached (see Sect. 10.3.2).

The conducted noise was studied obtaining the signal from the trigger path, i.e. by performing a rate scan at the output of the L1 mezzanine. The rate is the number of events that triggered within a pre-determined time window at a certain DT level. In our study, the width of the time window is 10 ms. We change the DT value from the maximum value at 1.2–0 V (negative values are not possible). This decrease is done with two different steps: one of them, the so-called coarse scan, follows steps of 4.8 mV while a more precise one, the fine scan, decreases in two steps of 1.2 mV and one of 2.4 mV.

This rate scan test was carried out over 5 different L0+L1 mezzanines. In Figs. 10.11, 10.12, 10.13, 10.14 and 10.15, we compare the results of the 5 mezzanines applying both coarse and fine scans when using the CTA PSU and linear PSU to power the *Dragon* Board. The following results were obtained when no PMT was attached to the Slow Control board and using the maximum DT L0 value to avoid any incoming signal. For convenience, in Table 10.2, the results of each configuration are summarized.

There are several things to take into account on these results. First of all, there is only one peak in each case which corresponds to the trigger of the baseline line. The existence of only one peak implies that no further noise/signal, besides the baseline, triggers. This peak has consistent position in both coarse and fine scans when using different PSU. We can appreciate that it does not peak at 0 mV in all measurements as could be expected. This is due to an offset produced by the electronics (mainly from the ASIC) that will be calibrated during future camera calibration and that does not affect our noise results. Nevertheless, the position of the peak is slightly shifted when the CTA PSU powers the system, which means that a systematic increase of the baseline is produced. Although the origin of this

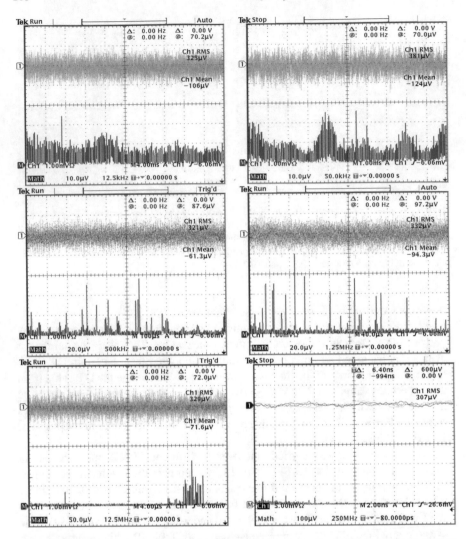

Fig. 10.10 Voltage versus time at different sampling frequency levels of 125 kHz (*top left*), 500 kHz (*top right*), 5 MHz (*middle left*), 12.5 MHz (*middle right*), 125 MHz (*bottom left*) and 2.5 GHz (*bottom right*) to look for high frequency noise is shown in the upper part. The vertical scale is 1 mV/div for all the plots except for the last one, at 2.5 GHz, whose scale is 5 mV/div. With a solid line, at the bottom of each plot, the FFT is shown with a vertical scale of 10 μV/div, 10 μV/div, 20 μV/div, 20 μV/div, 50 μV/div and 100 μV/div, respectively. For these measurements, all the involved devices were turned on (see Sects. 10.3.2 and 10.3.3)

increase is not understood, its value is very low and always below 15 mV (see Table 10.2).

Another important parameter is the width of the peak that provides information on the amplitude of the noise. In all cases, the width of the peak is narrow, which

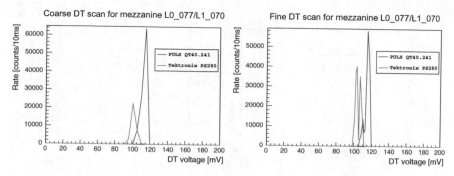

Fig. 10.11 L1 DT rate scan for the L0+L1 mezzanine L0 077–L1 070. The vertical axis shows rate (counts/10 ms) and X-axis DT voltage in mV. The blue line corresponds to the results from the PULS QT40.241 PSU, the brown one displays linear PSU signal. The leftmost plot shows the results when a coarse DT scan was performed while the one on the right panel shows the fine scan

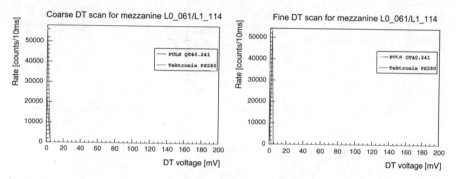

Fig. 10.12 Same as Fig. 10.11, but for L0+L1 mezzanine L0 061–L1 114

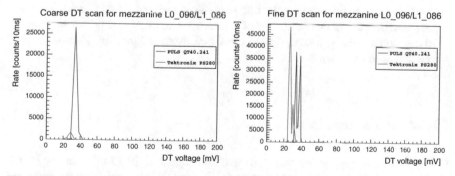

Fig. 10.13 Same as Fig. 10.11, but for L0+L1 mezzanine L0 096–L1 086

means that the baseline is confined to a short DT range. There are some small differences between the peak's width obtained while powering with the CTA PSU or linear PSU from one mezzanine to another, but the behavior is not constant or systematic and the difference is always lower than 5 mV. This fact allows us to

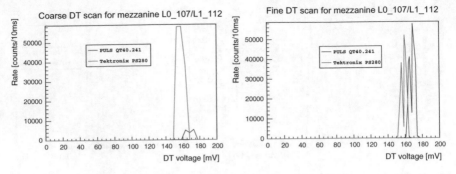

Fig. 10.14 Same as Fig. 10.11, but for L0+L1 mezzanine L0 107–L1 112

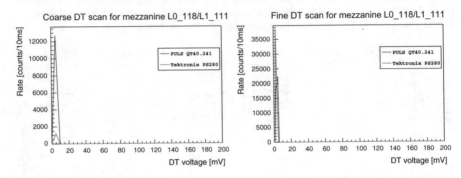

Fig. 10.15 Same as Fig. 10.11, but for L0+L1 mezzanine L0 118–L1 111

confirm that there is no evidence of a noise increase when we use the CTA PSU. In some cases, mainly during the fine scan, we could find double peaks. They are produced because of an overloaded of the rate in an amount of time shorter than the 10 ms. Consequently, the rate is reset to 0 before the integration window finishes giving rise to these structures.

10.3.4   Radiative Noise Tests

In Fig. 10.16 the comparison between the output signal of the PMT when the PULS QT40.241 PSU, placed nearby, is powered OFF and ON is shown. We can appreciate that no significant increase of the noise is induced when the PSU is working alongside.

Table 10.2 Conducted noise results for 5 L0+L1 mezzanines. The width of the peak (in mV) as well as the DT voltage at which the maximum rate is obtained (also in mV) is shown for the two different rate scans, coarse and fine, that differed on the DT steps (4.8 mV in the former and two steps of 1.2 mV and one of 2.4 mV in the latter). The DT voltage with the maximum rate provides an estimation on the peak position

		Coarse scan		Fine scan	
		Width [mV]	Voltage at max. rate [mV]	Width [mV]	Voltage at max. rate [mV]
$1L_0$ 077–L_1 070	PULS QT40.421	19.2	115.2	20.4	115.2
	Tektronix PS280	14.4	100.8	16.8	103.2
	Difference	4.8	14.4	3.6	12.0
$1L_0$ 061–L_1 114	PULS QT40.421	0	0	4.8	2.4
	Tektronix PS280	0	0	2.4	0
	Difference	0	0	2.4	2.4
$1L_0$ 096–L_1 086	PULS QT40.421	9.6	33.6	10.8	33.6
	Tektronix PS280	4.8	28.8	9.6	26.4
	Difference	4.8	4.8	1.2	7.2
$1L_0$ 107 L_1 112	PULS QT40.421	14.4	172.8	16.8	169.2
	Tektronix PS280	19.2	158.4	19.2	159.6
	Difference	−4.8	14.4	−2.4	9.6
$1L_0$ 118 L_1 111	PULS QT40.421	0	4.8	3.6	4.8
	Tektronix PS280	0	4.8	3.6	2.4
	Difference	0	0	0	2.4

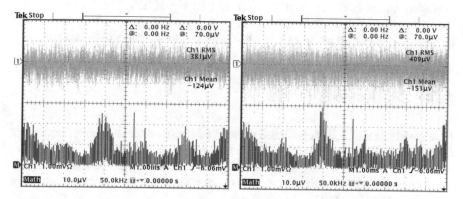

Fig. 10.16 Voltage versus time for a sampling frequency of 500 kHz. The vertical scale is 1 mV/div. With a solid line, at the bottom of each plot, the FFT is shown with a vertical scale of 10 μV/div. For both measurement, the PMT was powered by the linear PSU. On the left panel, the results obtained with the PULS QT40.241 PSU powered off are shown. On the right panel, results with all the involved devices turned on are depicted (see Sects. 10.3.2 and 10.3.3)

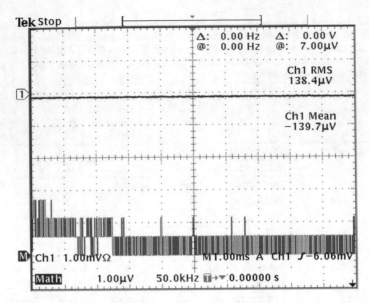

Fig. 10.17 Voltage versus time with 512 sample averaging at a frequency of 500 kHz. The vertical scale is 1 mV/div. With a solid line, at the bottom of each plot, the FFT is shown with a vertical scale of 1 μV/div. For this measurement, all the involved devices were turned on (see Sects. 10.3.2 and 10.3.3)

To improve the estimation of the radiated noise, we also obtained a 512 sample averaging signal in the range of 0–500 kHz (see Fig. 10.17). The Y-axis scale for the FFT is 1 μV/div. Therefore, the average noise is less than 2 μV at most of the frequencies (the noise increases slightly up to 7 μV in the range of 0–50 kHz).

10.4   Conclusions

No appreciable radiated or conducted noise was detected at the switching frequencies of the PULS QT40.241 PSU. No significant noise was detected using the linear PSU either. The conducted noise through the Dragon Board's L1 trigger path was analyzed obtaining satisfactory results that yielded to an absence of noise increase when using the PULS QT40.241 PSU.

Part V
Summary and Conclusion

Chapter 11
Summary and Concluding Remarks

During my thesis, I studied three of the best microquasar candidates to emit at VHE: Cygnus X-1, Cygnus X-3 and V404 Cygni. For the first time, we firmly detected Cygnus X-1 in the HE regime owing to the analysis of the 7.5 years of pass 8 *Fermi-LAT* data. This analysis allowed us to determine that HE gamma-ray emission only occurs during the HS and most likely associated to the relativistic jets. In turn, we found a hint of orbital modulation, which could constrain the gamma-ray emission site between $10^{11}-10^{13}$ cm from the BH produced by anisotropic IC on the stellar photon field. The study of this source is complemented by a long-term campaign performed with the MAGIC telescopes for a total of ~100 h. No detection was achieved at VHE, which allowed us to discard the jet-medium interaction region as possible production site for VHE gamma rays at the level of MAGIC sensitivity. On the other hand, we observed Cygnus X-3 with MAGIC for a total of ~70 h, covering a radio and HE flare from the beginning until its cessation, whose characteristics at those wavelengths were similar to the outbursts at which the system was detected in the HE regime in 2009. We did not detect the system, possibly related to the unusual high absorption produced by its companion star. Thus, VHE emission, if any, should be produced inside the binary system ($<10^{13}$ cm from the compact object). Thanks to this deep campaign, we showed that even more than 50 h during flaring activity would be needed with the future more sensitive CTA instrument. Finally, we observed the low-mass microquasar V404 Cygni during its outburst period in June 2015, produced after more than 25 years in quiescent state. Observations were carried out during strong hard X-ray flares and 1 h simultaneously to a hint in HE. However, detection was not achieved despite the low absorption displayed by the system at distances $>10^{10}$ cm from the BH, even during flaring activity. Thus, if the VHE emitter is located in the same region where HE gamma rays are produced, our non-detection would imply inefficient particle acceleration inside V404 Cygni jets or not enough energetics.

© Springer Nature Switzerland AG 2018
A. Fernández Barral, *Extreme Particle Acceleration in Microquasar Jets and Pulsar Wind Nebulae with the MAGIC Telescopes*, Springer Theses,
https://doi.org/10.1007/978-3-319-97538-2_11

In this thesis, it is also shown the first joint work between HAWC Observatory, *Fermi-LAT* and MAGIC, showing the results for a follow-up studies on detected sources by HAWC. We investigated the possible PWN nature of two sources, which allowed us to discard any relation between nearby pulsars and the HAWC candidates. The MAGIC analyses of these two sources were performed assuming both point-like and extended hypotheses, which provided constraints on a possible extension, likely higher than 0.16° radius. Part of my thesis is also dedicated to the study of PWNe. Here I analyzed five promising PWN candidates, given their hosting pulsars' features. These observations with MAGIC yielded no detection or hint of VHE emission. I set these results in context with the deep TeV PWN population study performed by the H.E.S.S Collaboration, concluding that our candidates are not outliers with respect to the detected PWNe and hence, gamma-ray emission could be expected. Delving into possible reasons for a non-detection, we conclude that these five PWNe are limited by the surrounding low target photon field, which turns the IC mechanism inefficient. Thus, we provide a general relation between the target IR photon field and the spin-down power of the pulsars hosted by detected and non-detected PWNe.

I included in my thesis the first VHE gamma-ray results for a Type Ia SN, SN 2014J. No significant excess was found during the first days after the explosion, which limited the total energy emitted in VHE gamma rays at $<10^{45}$ erg, which is about 10^{-6} of the total available energy budget of the SN explosion. Making use of a time-dependent flux model for hadronic origin and under certain assumptions, we proved that a power-law density profile is consistent with our results (although more sophisticated theoretical scenarios could shed more light). Following this result, gamma-ray detection from SN 2014J is not expected by any current of future generation of IACTs.

Regarding the technical part of my thesis, I worked in the QC tests of several subsystems for the camera of the future LST, mainly on PSU and trigger mezzanines. With dedicated conductive and radiate noise tests, were were able to prove the good functionality of the PSUs and allowed us also to approve the use of the proposed switching PSU for the LST.

Appendix A
Very-High-Energy Gamma-Ray Observations of SN 2014J

A.1 The Death of a Star: Types of Supernovae

As shown in Sect. 3.1, after a star reaches a critical mass, known as Chandrasekhar mass ($\sim 1.4 M_\odot$; Chandrasekhar 1931), an inevitable explosion that leads to its death takes place. This explosion is the so-called SN that releases an extremely amount of energy, around 10^{51} erg (Bethe 1993). All this energy, emitted in an unique explosion, corresponds to the totality of energy produced by our Sun in its entire life.

There are several types of SNe, that can be classified according to the amount of H observed in their spectra as well as because of the differences highlighted in their light curves:

- **Type I SN**: The explosions which do not present H lines in their spectra are classified within this group. In turn, they can be subdivided into three types: *Type Ia, Ib* and *Ic*. The former displays a strong absorption line of ionized Si II (around the wavelength 6510 Å) close to the maximum. Type Ib SNe lack of such spectral feature and are characterized by the presence of strong He lines absorption, while Type Ic do not show either Si or He absorption in their spectra (see Fig. A.1).
- **Type II SN**: Contrary to their counterparts, this type of explosions presents H line in the spectrum. They can be divided according to their light curves: classified as *Type II-L*, if after the maximum the luminosity decays linearly, or as *Type II-P*, if it remains bright (on a *plateau*) for a few months after reaching the highest luminosity.

Nevertheless, apart from this standard classification, the progenitor, source of energy and remnant of Type Ib and Type Ic SN are the same as for Type II SNe, evidencing a clear difference between Type Ia SNe and their correlative. As a general view, Type II (and consequently, Type Ib and Ic SNe), originate from the collapse of a massive star once the gravitational force cannot be handle by nuclear reaction in the core: Type Ib loose their H-rich outer layer, revealing the He-rich layer below, whilst Type Ic suffer more mass loss and looses both layers (in all cases, the collapse

© Springer Nature Switzerland AG 2018
A. Fernández Barral, *Extreme Particle Acceleration in Microquasar Jets and Pulsar Wind Nebulae with the MAGIC Telescopes*, Springer Theses,
https://doi.org/10.1007/978-3-319-97538-2

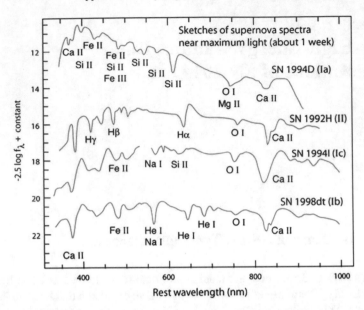

Fig. A.1 Sketches of SNe spectra, where the differences on the absorption lines are evidenced between each type (Carroll et al. 2006, data attributed to Thomas Matheson of National Optical Astronomy Observatory)

of an iron core is produced and the differences only concern the layers). All of them leave behind a compact object (either a NS or a BH).

On the other hand, Type Ia SNe originate from the detonation of a primary WD in a binary system that overpasses the Chandrasekhar mass limit. Unlike novae,[1] the Type Ia SNe are not recurrent phenomena: the nuclear energy released in such thermonuclear explosion is that high that no compact remnant is left behind.

There are three types of WDs: He, O-Ne and C-O WDs. The former can be excluded as progenitor of Type Ia SNe as their maximum mass is $\sim$0.45 $M_\odot$, otherwise He would start being burnt in the core. The second type is also excluded because they show C lines in their spectra (which is not a usual feature of these SNe). Therefore, it is believed that the main star of the binary that leads to these thermonuclear explosions are C-O WDs.

The nature of the companion star is still unclear, although two classical scenarios have been promoted: single-degenerate model, in which the WD accretes material from a red giant star (Whelan et al. 1973), and the double-degenerate model, in which the explosion is produced by the merging of two WDs (Iben et al. 1984).

- **Single-degenerate model**: The WD accretes material from the red giant via Roche lobe overflow or as wind-driven form given the strong wind of the giant companion. This material is H-rich that burns into He and afterwards into C increasing the mass

---

[1]**Novae** are abrupt increases of the luminosity on a WD binary system in a very short period time ($\sim$ days), which can be recurrent.

of the WD progressively. However, the accretion rate plays an important role: if it is too low, the H will be burnt and expelled in the surface giving rise to classical novae; if it is too high, the C will fuse and convert the star into an O-Ne WD, which are expected to die as Type II SNe (Saio et al. 1985). This delicate constraint on the accretion rate decreases the expected number of SNe arising from this model.

- **Double-degenerate model**: In this scenario, both components of the system are WDs. In this case, the transfer of material does not start due to evolutionary effect but because gravitational energy loss, leading the orbit that the objects follow to shrink and hence, to make them approach each other. Thus, once one of the WDs fills its Roche lobe, the mass transfer begins. As before, the merger rate seems to be insufficient to explain all the Type Ia SNe.

Still, the evolutionary path that leads to a C-O WD which exceeds the Chandrasekhar limit is not well understood yet. Moreover, neither the conditions for the successful detonation by assuming any of the scenarios is well-understood. Nevertheless, despite these difficulties, one of the most remarkable features of the subclass Type Ia SNe is that all of them present approximately the same luminosity peak and decay slope, which convert these systems into very interesting objects for cosmological measurements (see, e.g. Carroll et al. 2006), as the accelerated expansion of our Universe (Perlmutter et al. 1999) or on Galactic chemical evolution (Timmes et al. 1995).

All the ejected material by any SN expands and interacts with the ISM, giving rise to a new astrophysical source known as SNR. SNRs experiment different phases along their evolution. However, the aim of this Appendix is to focus on the results of SN 2014J, a Type Ia SN observed by MAGIC at its very early stage and so, in the following I will only describe the first phase of a SNR. For a more comprehensive and detailed information on this type of sources, the reader is referred to Reynolds (2008).

The first stage of a SNR evolutionary path is the so-call **free expansion phase**, in which the ejected material expands without being decelerated. This phase is independent of the nature of the SN (Type I or II), since the energy and density of the ejected material always exceed the values of the surrounding medium right after the explosion. The shock wave created by the explosion moves within the more or less homogeneous interstellar gas at supersonic velocities ($v_{ej} \sim 10^4$ km/s). Although the ISM can be assumed roughly homogeneous, this is not always true: progenitors of Type II and single-degenerate Type Ia SNe might create surrounding circumstellar medium (with a density profile of $\rho \propto r^{-2}$) due to their strong stellar wind in which the SNR evolves. However, small scales inhomogeneities won't affect the structure of the ejecta. Only major inhomogeneities, like circumstellar medium in equatorial disks or jet-driven SNe, can disturb the symmetry, which are not typical and hence, homogeneous medium are normally assumed in the discussion of SNRs at early phases. Based on this assumption then, during this first stage, the ISM does not influence in the expansion of the shock front, i.e. the pressure applied by the ISM is negligible. As the strong shock front expands, the interstellar gas accumulates, separated from the ejected material by the so-called *contact discontinuity*.[2] At this point, a reverse shock starts forming behind this *contact discontinuity*. The classical criterion that terminates this phase is the equality between the mass of the ISM (compressed between the shock front and the *contact discontinuity*) and the initial ejected stellar mass (M_{ej}). When this happens the SNR displays the denominated sweep-up radius, R_{sw}, which depends on the initial density of the ISM, n_0:

$$M_{ej} = \frac{4\pi}{3} \left(R_{sw}^3 n_0 \right) \Rightarrow R_{sw} = \left(\frac{3M_{ej}}{4\pi n_0} \right)^{1/3} \tag{A.1}$$

This radius is reached at a time of $t_{sw} = R_{sw}/v_{ej}$, that can expand hundred of years depending on the surrounding medium. At the end of this phase, the reverse shock accumulates enough mass and starts moving into the opposite direction of the ejected material. This inward movement heats the ejected material to high temperatures, a flat pressure structure is developed and therefore, the expansion of the SNR is produced by the thermal pressure of the hot gas. This is the end of the first stage and the beginning of the so-called Sedov-Taylor phase.

A.2 Introduction to SN 2014J

On the 21st of January 2014 (MJD 56678), SN 2014J was detected by the University College London (UCL) Observatory (Fossey et al. 2014) and classified as a Type Ia SN with the Dual Imaging Spectrograph on the Astrophysical Research Consortium (ARC) 3.5 m telescope (22nd of January; Goobar et al. 2014). It is located in the starburst galaxy M82 at a distance of 3.6 Mpc (Karachentsev et al. 2006). Its proximity has granted it the title of the nearest Type Ia SN since 1972 and motivated large multiwavelength follow-up observations from radio to VHE gamma rays.

Deep studies of color excess and reddening estimation were carried out on SN 2014J, phenomena associated to the interstellar extinction of the radiation due to its absorption or scattering by gas or dust in the medium. Type Ia SNe represent good candidates for reddening studies given their very high luminosity and similarity from one to another. Amanullah et al. (2014) reported, for the first time, a characterization of the reddening of a Type Ia SN in a full range from 0.2 to 2 µm. Their results, with reddening values of $E_{B-V} \sim 1.3$ and $R_V \sim 1.4$, are compatible with a power-law extinction, expected in the case of multiple scattering scenarios. In the same wavelength band, from UV to NIR, Foley et al. (2014) found reddening parameter values of $E_{B-V} \sim 1.2$ and $R_V \sim 1.4$. In this case, the extinction is explained to be caused by a combination of the galaxy dust and a dusty circumstellar medium. However, although compatible with the former extinction law (at a low value of $R_V \sim 1.4$) and consistent as well with previously mentioned results, Brown et al. (2015),

---

[2]The **contact discontinuity** is the the surface between two different materials with similar pressure and velocities but different densities.

making use of *Swift*-UVOT data, suggested that most of the reddening is caused by the interstellar dust. Optical and NIR linear polimetric observations of the source presented in Kawabata et al. (2014) supports the scenario where the extinction is mostly produced by the interstellar dust. These evidences favor **the double-degenerate scenario** for SN 2014J, where less circumstellar dust is expected than in cases with a giant companion star. This type of companions is indeed ruled out by several authors as possible progenitors in SN 2014J, e.g. Pérez-Torres et al. (2014), with the most sensitive study of a Type Ia SN in the radio band, or Margutti et al. (2014), in the X-ray band. The former reported non-detection from the observations performed with eMERLIN and EVN. These results, compared with detailed modeling of the radio emission from the source, allowed them to exclude the single-degenerate scenario in favour of the double-degenerate one with constant density medium of $n \lesssim 1.3\,\mathrm{cm}^{-3}$.

Several authors have speculated about the possibility of SN explosions being able to produce gamma-ray emission at detectable level by current and/or future telescopes. This gamma-ray emission is associated to the diffuse particle acceleration that the strong shock fronts of extreme detonations like SNe produce. However, these models generally consider Type II SNe due to the strong wind of the progenitors which provide larger amount of targets (e.g. Kirk et al. 1995 and Tatischeff 2009). Nevertheless, given the proximity of SN 2014J, this event provides a good exploratory opportunity to probe the eventual production of VHE gamma rays during the first days after such an explosion.

A.3   MAGIC Observations and Results

SN 2014J was observed with the MAGIC telescopes under moderate moonlight conditions from the 27th to the 29th of January and on the 1st and 2nd of February of 2014 under dark-night conditions at medium zenith angles (from $40°$ to $52°$). Our observations, performed using the *wobble-mode* (see Sect. 2.4.2.1), started six days after the first detection by the UCL Observatory because of adverse weather conditions. The complete data set up to $50°$ ($\sim$5.5 h) was used for the analysis given the overall good quality of the data (concerning weather, light conditions and performance of the system).

The analysis was performed using the analysis pipeline described in Sect. 2.4.3. Figure A.2 shows the θ^2 distribution, i.e the squared angular distance between the reconstructed gamma-ray direction and the position of either SN 2014J (on-source histogram) or the center of the background control region (off-source histogram). Standard LE cuts were used, which means that a selection of $\theta^2 < 0.02\,\mathrm{deg}^2$, $hadronness < 0.28$ and $size > 60$ phe (in both telescopes) was applied. The resulting excess of the on-source histogram over the background from the region, where gamma-ray events from SN 2014J are expected, is compatible with background. The significance computed using Eq. 2.12 is 0.90σ.

ULs on the flux were computed for 95% C.L., assuming a power-law spectrum, $dF/dE \propto E^{-\Gamma}$, with photon index of 2.6. Variations of $\sim$20% in the photon index

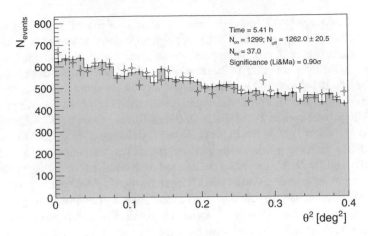

Fig. A.2 Distribution of the squared angular distance, θ^2, after 5.41 h of observation between the reconstructed arrival direction of the gamma-ray candidate events and the position of the source in the camera (red empty circles). The θ^2 distribution of the background events (black points) is also displayed. The vertical dashed line at $\theta^2 = 0.02$ deg^2 defines the expected signal region. Credit: Ahnen et al. (2017), reproduced with permission © ESO

Table A.1 Summary of the MAGIC observations of SN 2014J. *From left to right*: date of the beginning of the observations, also in MJD, effective time, zenith angle range and integral ULs at 95% C.L. above 300 and 700 GeV. The last row reports the integral ULs derived with the entire data sample. Due to low statistics, no integral UL was computed for energies above 700 GeV for the first day of observations. Credit: Ahnen et al. (2017), reproduced with permission ©ESO

Date		Eff. Time [hours]	Zd [°]	UL ($E > 300$ GeV) [photons cm^{-2} s^{-1}]	UL ($E > 700$ GeV) [photons cm^{-2} s^{-1}]
[yyyy-mm-dd]	[MJD]				
2014-01-27	56684.23	0.43	47–50	1.03×10^{-11}	–
2014-01-28	56685.06	1.41	40–43	2.19×10^{-12}	1.55×10^{-12}
2014-01-29	56686.09	1.30	40–42	4.55×10^{-12}	5.97×10^{-13}
2014-02-01	56689.07	0.98	40–42	3.14×10^{-12}	9.98×10^{-13}
2014-02-02	56690.08	1.30	40–42	3.35×10^{-12}	1.76×10^{-12}
Total	–	5.41	40–50	1.30×10^{-12}	4.10×10^{-13}

produced changes in the integral ULs of less than 5%, and hence small deviations from the used value do not critically affect the reported ULs. The ULs above 300 and 700 GeV for the single-night observations are reported in Table A.1 and depicted in Fig. A.3.

After $\sim$5.5 h of observations with the MAGIC telescopes, we establish an integral UL on the gamma-ray flux for energies above 300 GeV of 1.3×10^{-12} photons cm^{-2}s^{-1} at 95% C.L., which corresponds to 1.0 % C.U. in the same energy range. For energies above 700 GeV, the integral UL is 4.1×10^{-13} photons cm^{-2}s^{-1}, corresponding to 1.1 % C.U. at the same C.L. Our ULs for $E > 700$ GeV are already

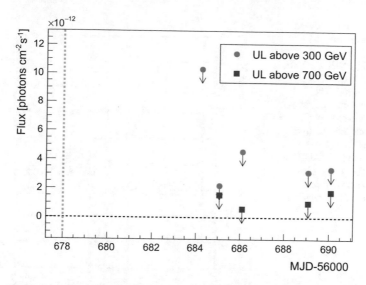

Fig. A.3 MAGIC daily integral ULs from the direction of SN 2014J for energies above 300 GeV (circles) and 700 GeV (squares). The integral UL for energies above 700 GeV was not computed for the first night (MJD 56684) due to low statistics (see also Table A.1). The horizontal dashed line indicates zero flux level and the vertical dashed line indicates the day of the SN explosion (MJD 56678), just six days before the beginning of the MAGIC observations. Credit: Ahnen et al. (2017), reproduced with permission © ESO

close to the flux from the host galaxy M82 measured by VERITAS in the same energy range, $(3.7 \pm 0.8_{stat} \pm 0.7_{syst}) \times 10^{-13}$ photons cm^{-2} s^{-1} (VERITAS Collaboration et al. 2009), which constitutes an irreducible background for our measurement. Under the hypothesis that M82 has a gamma-ray spectrum of $dF/dE = 3 \times 10^{-16}(E/1000$ GeV$)^{-2.5}$ photons cm^{-2} s^{-1} GeV^{-1}, as measured by VERITAS, the expected number of excess events in our observations would be 9.4, with a 95% C.L. lower limit at -6.1 (obtained by means of the full likelihood method, see Sect. 2.4.3.13). The observed number of excess events by MAGIC is -4.2, with an associated p-value of 8.4×10^{-2}, hence consistent with the VHE flux of M82 measured by VERITAS (see Fig. A.4).

A.4   Discussion

As mentioned before, VHE gamma-ray emission can be expected from the SN explosion and its remnant by the strong shock front produced. In the literature, both hadronic and leptonic origin have been discussed (Aharonian 2013). In the former, gamma rays result from the decay of neutral pions, π^0, as a consequence of inelastic collisions between the protons accelerated in the SN and the ambient atomic nuclei. In the leptonic scenario, the most efficient mechanism to radiate VHE gamma rays

Fig. A.4 Full likelihood
output for the SN 2014J,
assuming the spectral shape
measured by VERITAS for
the host galaxy M82 and the
IRF of MAGIC. The
expected number of excess
9.4 has a lower 95% C.L. of
−6.1 events, compatible
with the excess of −4.2
events found by MAGIC at
energies above 700 GeV,
making VERITAS and
MAGIC results consistent to
each other

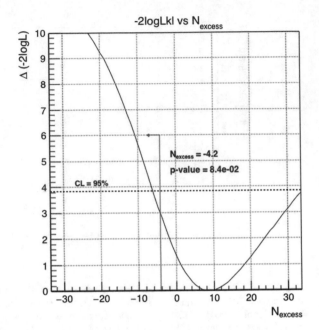

seems to be the IC process of accelerated electrons on ambient photons. In both
cases, the environment plays an important role for the production of VHE gamma-
ray radiation. A near and young supernova (∼1 week old) emitting in this energy
regime could shed light on the progenitors of these thermonuclear stellar explosions.

Although we did not detect VHE gamma rays right after the explosion, using the
known distance of M82 ($d_{M82} = 3.6$ Mpc Karachentsev et al. 2006) and assuming,
as before, a photon index of 2.6, one can convert the measured flux UL into an UL
on the power emitted into VHE gamma rays. Therefore, given the integral UL for
energies greater than 300 GeV, 1.3×10^{-12} photons cm^{-2}s^{-1}, the resulting UL on
the power emitted is of the order of 10^{39} erg s^{-1}. If one now assumes an emission
period of the order of 10 days, the total energy emitted in VHE gamma rays during
this period is smaller than 10^{45} erg, which is about 10^{-6} of the total available energy
budget of the SN explosion (∼10^{51} erg, Bethe 1993).

Models of the evolution of young SNRs can be used to estimate the expected
emission from the region in the future. One of the most important parameters to be
assumed is the density profile of the SN ejecta. In this work, we considered a simple
power-law density profile, which allow us to use the Dwarkadas (2013) model to
obtain an analytic solution for the estimated flux. Other density profiles have been
used in the literature: Models like W7 or WDD1, applied by Nomoto et al. (1984)
and Iwamoto et al. (1999), are usually utilized in Type Ia SN studies, but they are
based in the single-degenerate scenario. Dwarkadas et al. (1998) discussed a possible
exponential density profile which could represent better the SN ejecta structure than
the power-law one. However, the exponential profile cannot provide an analytic result

as the one assumed in this work, which can give a correct solution within the order of magnitude, as explained in Dwarkadas (2013).

Thus, making use of equation 10 in Dwarkadas (2013), one can obtain the time-dependent emission assuming a hadronic origin. Considering only this hadronic origin, we can establish a lower limit on the total gamma-ray radiation. As discussed above, gamma-ray emission can be also expected from IC processes. Nevertheless, purely leptonic scenario have been questioned and discarded by several authors, e.g. Völk et al. (2008).

The expected flux depends strongly not only on the assumed density structure of the SNR but also on the density profile of the surrounding ISM. As shown by different authors, we can consider double-degenerate scenario in the case of SN 2014J, i.e. two WDs progenitors. WDs do not suffer wind-driven mass-loss and therefore are not expected to modify the surrounding medium, although different assumptions, from the lack of certainty on the progenitors, have also been studied (see e.g. Dwarkadas 2000). We can then assume that the Type Ia SN explosion took place in a constant density medium. In this work, we used a density of $n = 2.2 \times 10^{-24}$ g/cm^3 (Pérez-Torres et al. 2014), assuming that all the content in the host galaxy of the source, M82, stems from neutral hydrogen, H_I. This homogeneous medium assumption leads to an increasing flux emission, above a certain gamma-ray energy, with time in the free-expansion SNR stage, as shown below in the expression given by Dwarkadas (2013):

$$F_\gamma(> 1\text{TeV}, t) = \frac{3q_\gamma \xi (\kappa C_1)^5 m^3}{6(5m-2)\beta \mu m_p d^2} n^2 t^{5m-2} \qquad (A.2)$$

where the assumed parameters in this work are

- $q_\gamma = 1 \times 10^{-19}$ cm^{3}s^{-1}erg^{-1}H-atom^{-1} (for energies greater than 1 TeV) is the emissivity of gamma rays normalised to the cosmic ray energy density tabulated in Drury et al. (1994). This value corresponds to a spectral index of 4.6 of the parent cosmic ray distribution, which was selected according to the assumed spectral index in this work, $\Gamma = 2.6$;
- $\xi = 0.1$ is the fraction of the total SN explosion energy converted to cosmic ray energy, so an efficient cosmic ray acceleration is assumed;
- $\kappa = 1.2$ is the ratio between the radius of the forward shock and the contact discontinuity (which separates ejecta and reverse shock);
- $C_1 = 1.25 \times 10^{13}$cm/s^m is referred to as a constant related to the kinematics of the SN. This value is calculated from the relation given by Dwarkadas (2013), $R_{shock} = \kappa C_1 t^m$. In turn, R_{shock} is obtained from equation 2 in Gabici et al. (2016), by assuming an explosion energy of 10^{51} erg, a mass of the ejecta of 1.4 $M_\odot$ and a ISM density of 1.3 cm^{-3}, whose value is constrained by Pérez-Torres et al. (2014);
- $\beta = 0.5$ represents the volume fraction of the already shocked region from which the emission arises;
- $\mu = 1.4$ is the mean molecular weight;
- $m_p = 1.6 \times 10^{-24}$g is the proton mass;
- $d = 3.6$ Mpc is the distance to our source;

- t is the elapsed time since the explosion; and
- m is the expansion parameter.

The expansion parameter varies along the free-expansion phase in different ways according to the assumed model for the density structure of the SN ejecta after the explosion. For the discussion of this source, I make use of the power-law profile with a density proportional to R^{-7} (Chevalier 1982), where R is the outer radius of the ejecta. The initial value of the expansion parameter is very unalike depending on the density profile assumed, but in all cases evolve to $m = 0.40$ (Dwarkadas et al. 1998). This limit at 0.40 is constrained by the beginning of the Sedov-Taylor phase.

The expansion parameter for the power-law profile keeps constant at 0.57 in the first years of the free-expansion stage. Given this value, the expected flux above 1 TeV (constrained by the emissivity of gamma rays, q_γ, tabulated in Drury et al. 1994) at the time of the MAGIC observations ($t = 6$ days) from Eq. A.2 is $\approx 10^{-24}$ photons cm^{-2}s^{-1}. This flux is consistent with the UL at 95% C.L. derived from MAGIC data in the same energy range, 2.8×10^{-13} photons cm^{-2}s^{-1} and hence, the power-law density profile could be considered as possible model to describe the density structure of SN 2014J, considering all the assumptions and parameters selection discussed above. On the other hand, this model predicts a constant parameter of $m = 0.57$ during the first ~ 300 years, after which it starts dropping gradually (Dwarkadas et al. 1998). Although the flux keeps increasing with time according to Eq. A.2, with this low expansion parameter it will still be about 10^{-21} photons cm^{-2}s^{-1} 100 years after the SN occurred, which is well below the sensitivity of the current and planned VHE observatories.

A.5 Conclusions

MAGIC observed SN 2014J, the nearest Type Ia SN since 1972, just 6 days after the explosion. Its proximity offered a good chance to probe gamma-ray emission from this kind of source but no gamma-ray excess was found. Integral ULs for energies above 300 and 700 GeV were established at 1.3×10^{-12} photons cm^{-2}s^{-1} and 4.1×10^{-13} photons cm^{-2}s^{-1}, respectively, for a 95% C.L. and assuming a power-law spectrum. The latter are compatible with the results obtained by VERITAS during their observations on the host galaxy M82.

With the obtained flux UL, we were able to constrain the fraction of energy emitted into VHE gamma rays. Thus, the flux UL at $E > 300$ GeV corresponds to an emission power of $<10^{39}$ erg s^{-1} or a total maximal emitted VHE gamma-ray energy during the observational period—approximately ten days—of $<10^{45}$ erg, which is about 10^{-6} times the total energy budget of a Type Ia SN explosion ($\sim 10^{51}$ erg). Following Dwarkadas (2013) model for hadronic gamma-ray flux, a power-law density profile proportional to R^{-7} is consistent with our ULs, although, due to the uncertainties in several parameters, this cannot exclude other, more sophisticated, theoretical scenarios. Assuming this SN density profile and a constant density medium, we can estimate an expected emission from the region of the source of $\approx 10^{-25}$ photons $cm^{-2}s^{-1}$. Following these assumptions, this flux would not increase enough in a near future to be detectable by any current or future generation of IACTs.

Content included in this chapter has been published in Ahnen et al. (2017) (reproduced with permission © ESO).

Appendix B
F-Factor Method

The number of photons that hits the PMT photocathode follows a Poissonian distribution. This is produced by the fact that only a fraction of the incident photons are actually collected by the photocathode. The emission of phes through the photoelectric effect inside the dynode system is a random binary process. Thus, the number of phes obtained in the PMT would be given by the convolution of these two distributions, which leads to a Poissonian one:

$$P(n; N_{phe}) = \frac{N_{phe}^n}{n!} e^{-N_{phe}} \tag{B.1}$$

where $P(n; N_{phe})$ is the probability of observing n phes when the expected value is N_{phe}, defined as $N_{phe} = N_\gamma \cdot QE$ (where N_γ is the mean number of photons arriving to the photocathode and QE is the quantum efficiency of it). Therefore, the mean value is N_{phe} and the RMS is $\sqrt{N_{phe}}$.

On the other hand, the measured charge (Q) in ADC counts has a mean $\bar{Q}$ and a RMS of σ_Q (which is wider than a pure Poissonian RMS). Thus, the relation between both quantities can be expressed as:

$$F = \frac{1}{\sqrt{N_{phe}}} = \frac{\sigma_Q}{\bar{Q}} \Rightarrow N_{phe} = \left(\frac{\bar{Q}F}{\sigma_Q}\right)^2 \tag{B.2}$$

The F-factor, F, accounts for a broadening of the signal due to the multiplication process inside the dynode system. That is the reason why the F-factor is measured for each PMT separately before being installed in the telescope's camera. In MAGIC, this value is approximately 1.15 in all cases.

The three $\bar{Q}$, σ_Q (from the calibration events) and F-factor are known, which allows us to compute the mean N_{phe} from Eq. B.2 and to obtain, in turn, the conversion factor, C:

$$C = \frac{N_{phe}}{\bar{Q}} = \frac{F^2 \bar{Q}}{\sigma_Q^2} \tag{B.3}$$

© Springer Nature Switzerland AG 2018
A. Fernández Barral, *Extreme Particle Acceleration in Microquasar Jets and Pulsar Wind Nebulae with the MAGIC Telescopes*, Springer Theses,
https://doi.org/10.1007/978-3-319-97538-2

In MAGIC, the values of C (different for each PMT) are calculated with special calibration events. During data taking, more calibration flashes are fired to update them. Afterwards, these constantly updated conversion factors are applied to real data to get the number of phes from the charge, Q, of the ADC.

Appendix C
Power Supply Noise Tests with Different Setup Configurations

In this Appendix, I show plots from all configuration setups presented in Sect. 10.3 and not included there. For more information about the setup features for each case, please refer to the captions.

C.1 Conducted Noise Tests

Here I show only conducted noise tests using the linear PSU (Tektronix PS280). The data were recorded using the Tektronix TDS3024B oscilloscope (Figs. C.1 and C.2).

C.2 Radiated Noise Tests

For these tests, the power to the PMT (if any) was provided by the linear PSU while the PULS QT40.241 PSU was placed alongside. Data was recorded using the Tektronix TDS3024B oscilloscope. Figures C.3, C.4 and C.5 show the results of the tests with PULS QT40.241 PSU powered off and Figs. C.6 and C.7 with PULS QT40.241 PSU powered on (although unloaded).

© Springer Nature Switzerland AG 2018
A. Fernández Barral, *Extreme Particle Acceleration in Microquasar Jets and Pulsar Wind Nebulae with the MAGIC Telescopes*, Springer Theses,
https://doi.org/10.1007/978-3-319-97538-2

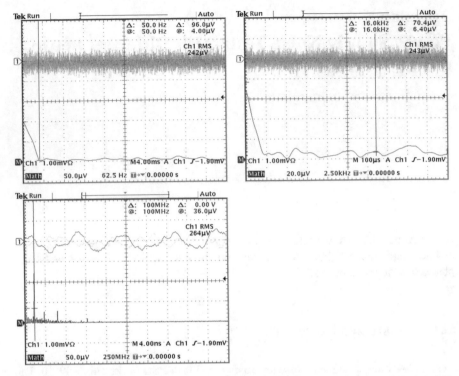

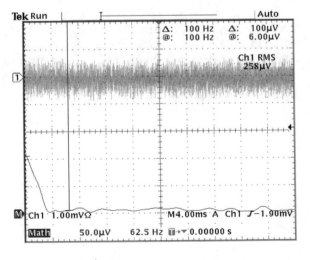

Fig. C.1 Voltage versus Time for the three different sampling frequency levels, 625 Hz (*top left*), 25 kHz (*top right*) and 2.50 GHz (*bottom left*). The vertical scale is 1 mV/div for all cases. With a solid line, at the bottom of each plot, the FFT is shown with a vertical scale of 50 μV/div, 20 μV/div and 50 μV/div, respectively. For these measurements, all involved devices were turned off (V_{cc} and HV off; see Sects. 10.3.0.2 and 10.3.0.3)

Fig. C.2 Voltage versus Time at sampling frequency of 625 Hz. The vertical scale is 1 mV/div. With a solid line, at the bottom of each plot, the FFT is shown with a vertical scale of 50 μV/div. For this measurement, only V_{cc} was powered on at the level of 5 V, while the HV remained powered off (see Sects. 10.3.0.2 and 10.3.0.3)

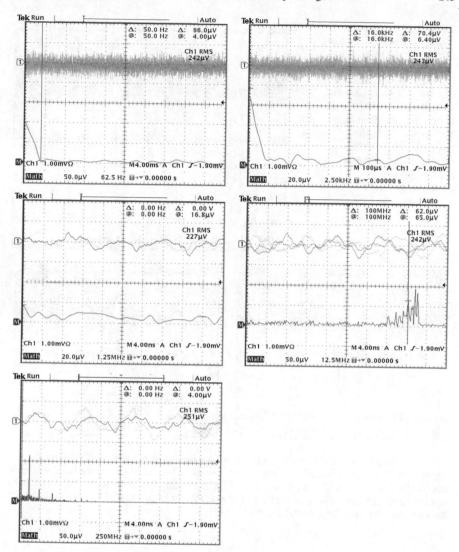

Fig. C.3 Voltage versus Time for five different sampling frequency levels, 625 Hz (*top left*), 25 kHz (*top right*), 12.5 MHz (*middle left*), 125 MHz (*middle right*) and 2.5 GHz (*bottom left*). The vertical scale is 1 mV/div for all cases. With a solid line, at the bottom of each plot, the FFT is shown with a vertical scale of 50 μV/div, 20 μV/div, 20 μV/div, 50 μV/div and 50 μV/div, respectively. For these measurements, all involved devices were turned off (PMT and PULS QT40.241 PSU off; see Sects. 10.3.0.2 and 10.3.0.3)

Fig. C.4 Voltage versus
Time at sampling frequency
of 625 Hz. The vertical scale
is 1 mV/div. With a solid
line, at the bottom of each
plot, the FFT is shown with a
vertical scale of 50 μV/div.
For this measurement, only
V_{cc} =5 V was supplied,
while the HV and PULS
QT40.241 PSU remained
powered off (see
Sects. 10.3.0.2 and 10.3.0.3)

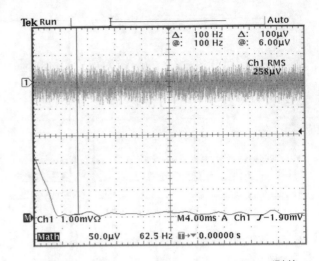

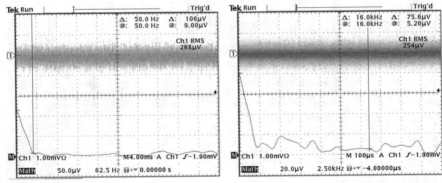

Fig. C.5 Voltage versus Time for the two different sampling frequency levels, 625 Hz and 25 kHz,
respectively. The vertical scale is 1 mV/div in both cases. With a solid line, at the bottom of each
plot, the FFT is shown with a vertical scale of 50 μV/div and 20 μV/div, respectively. For these
measurements, the linear PSU provided V_{cc}=5 V and HV=1.4 kV while the PULS QT40.241 PSU
was still turned off; see Sects. 10.3.0.2 and 10.3.0.3)

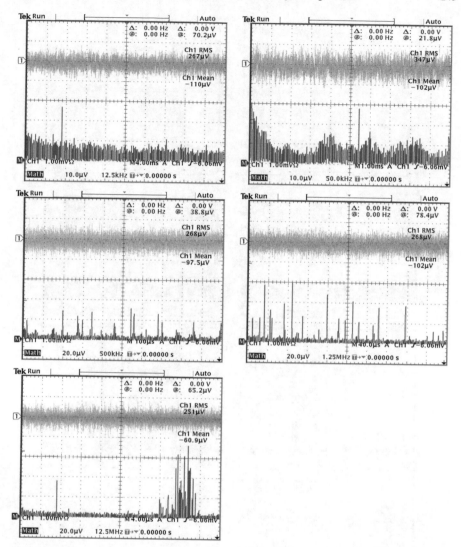

Fig. C.6 Voltage versus Time for five different sampling frequency levels, 125 kHz (*top left*), 500 kHz (*top right*), 5 MHz (*middle left*), 12.5 MHz (*middle right*) and 125 MHz (*bottom left*). The vertical scale is 1 mV/div for all cases. With a solid line, at the bottom of each plot, the FFT is shown with a vertical scale of 10 μV/div for the two top plots and 20 μV/div for the rest. For these measurements, V_{cc} and HV were powered off while the PULS QT40.241 PSU was on although unloaded; see Sects. 10.3.0.2 and 10.3.0.3)

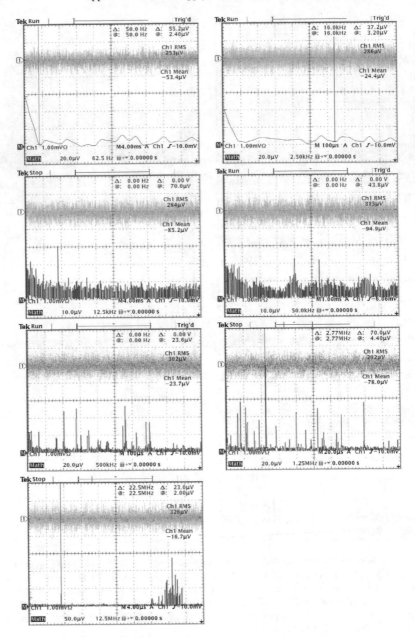

Fig. C.7 Voltage versus Time for seven different sampling frequency levels, 625 Hz, 25 kHz, 125 kHz, 500 kHz, 5 MHz, 12.5 MHz and 125 MHz, respectively. The vertical scale is 1 mV/div for all cases. With a solid line, at the bottom of each plot, the FFT is shown with a vertical scale of 20 μV/div, 20 μV/div, 10 μV/div, 10 μV/div, 20 μV/div, 20 μV/div and 50 μV/div, respectively. For these measurements, the PMT was powered on with V_{cc} = 5 V and HV = 1.4 kV, as well as the PULS QT40.241 PSU although unloaded; see Sects. 10.3.0.2 and 10.3.0.3)

References

Aharonian FA (2013) Astropart Phys 43:71
Ahnen ML et al (2017) A&A 602:A98
Amanullah R et al (2014) ApJ 788:L21
Bethe HA (1993) ApJ 412:192
Brown PJ et al (2015) ApJ 805:74
Carroll BW et al (2006) An introduction to modern astrophysics and cosmology
Chandrasekhar S (1931) ApJ 74:81
Chevalier RA (1982) ApJ 258:790
Drury LO et al (1994) Astron Astrophys 287:959
Dwarkadas VV et al (1998) ApJ 497:807
Dwarkadas VV (2000) ApJ 541:418
Dwarkadas VV (2013) Mon Not R Astron Soc 434:3368
Foley RJ et al (2014) MNRAS 443:2887
Fossey SJ et al (2014) Central Bureau Electronic Telegrams, 3792
Gabici S et al (2016) ArXiv e-prints
Goobar A et al (2014) ApJ 784:L12
Iben I Jr et al (1984) ApJS 54:335
Iwamoto K et al (1999) ApJS 125:439
Karachentsev ID et al (2006) Astrophysics 49:3
Kawabata KS et al (2014) ApJ 795:L4
Kirk JG et al (1995) A&A 293
Margutti R et al (2014) Astrophys J 790:52
Nomoto K et al (1984) ApJ 286:644
Pérez-Torres MA et al (2014) ApJ 792:38
Perlmutter S et al (1999) Astrophys J 517:565
Reynolds SP (2008) ARA&A 46:89
Saio H et al (1985) A&A 150:L21
Tatischeff V (2009) A&A 499:191
Timmes FX et al (1995) ApJS 98:617
VERITAS Collaboration et al (2009) Nature 462, 770
Völk HJ et al (2008) A&A 490:515
Whelan J et al (1973) ApJ 186:1007

Printed in the United States
By Bookmasters